Springer Theses

Recognizing Outstanding Ph.D. Research

Aims and Scope

The series "Springer Theses" brings together a selection of the very best Ph.D. theses from around the world and across the physical sciences. Each volume has been selected for its scientific excellence by an internationally renowned institute or department. For greater accessibility to non-specialists, the published versions include an extended introduction, as well as a foreword by the student's supervisor explaining the special relevance of the work for the field. The series as a whole provides a valuable resource both for newcomers to the research fields, and for other scientists seeking detailed background information on special questions. Finally, it serves as an accredited documentation of the valuable contributions made by today's younger generation of scientists.

How to Apply

Theses prepared at internationally leading universities and institutes may be nominated for publication by supervisors with the endorsement of the head of department. Each such department is entitled to nominate one thesis per year. Enquiries and requests for a nomination form should be sent, together with the thesis, to Ms. Nathalie Berg (see link on right).

Eligibility Criteria

To qualify for consideration, theses should fulfil the following criteria:

- They must be written in good English.
- The topic should fall within the confines of Chemistry, Physics, Earth Sciences, Engineering and related interdisciplinary fields such as Materials, Nanoscience, Chemical Engineering, Complex Systems and Biophysics.
- The work reported in the thesis must represent a significant scientific advance.
- A maximum of 30% of the text may be a verbatim reproduction from the author's previous publications.
- They must have been examined and passed during the 12 months prior to nomination.
- The theses should have a clearly defined structure including an introduction accessible to new PhD students and scientists not expert in the relevant field.
- After acceptance, theses should be supplemented by a "Supervisor's Foreword" outlining the wider scientific relevance of the work.

Indexed by zbMATH.

Garrett B. King

Electroweak Structure and Reactions in Light Nuclei with Quantum Monte Carlo Methods

Doctoral Thesis accepted by
Washington University, St. Louis, MO, USA

Springer

Author
Dr. Garrett B. King
Los Alamos National Laboratory
Los Alamos, NM, USA

Supervisors
Prof. Saori Pastore
Washington University
St. Louis, MO, USA

Prof. Maria Piarulli
Washington University
St. Louis, MO, USA

ISSN 2190-5053 ISSN 2190-5061 (electronic)
Springer Theses
ISBN 978-3-032-03190-7 ISBN 978-3-032-03191-4 (eBook)
https://doi.org/10.1007/978-3-032-03191-4

This Springer imprint is published by the registered company Springer Nature Switzerland AG
The registered company address is: Gewerbestrasse 11, 6330 Cham, Switzerland

Supervisors' Foreword

Garrett King's thesis focuses on theoretical nuclear physics, specifically using Quantum Monte Carlo computational methods to study nuclear electroweak observables. His research explores key topics such as beta-decay rates, muon capture, electromagnetic moments and form factors, and nuclear electroweak responses, providing theoretical predictions for both the Standard Model and searches for new physics. Garrett implemented models incorporating two- and three-nucleon interactions as well as many-nucleon electroweak currents thus contributing to precision tests of nuclear physics. He was the recipient of the 2025 American Physical Society—Division of Nuclear Physics Dissertation Award. Garrett is continuing his work at Los Alamos National Laboratory as a Director's Postdoctoral Fellow, further advancing his expertise in nuclear physics by studying electron- and neutrino-nucleus interactions to aid searches for fundamental symmetry research.

St. Louis, USA
July 2025

Prof. Saori Pastore
Prof. Maria Piarulli

Acknowledgements Many thanks and acknowledgements are in order for all of the support that I have received to make this thesis possible.

First and foremost, there would be no thesis without the guidance of my co-chairs, Saori Pastore and Maria Piarulli. I appreciate your support, encouragement, and graciousness with your time over the course of my thesis. A student would be lucky to have just one good advisor, but I am fortunate to have two great ones.

I am also lucky to have had a number of great mentors outside of Washington University. My sincerest gratitude is owed to Joe Carlson, Stefano Gandofli, Rocco Schiavilla, and Bob Wiringa for their help in answering myriad technical questions and for their scientific guidance.

To the members of my committee—Bhupal Dev, Willem Dickhoff, Kade Head-Marsden, and Lee Sobotka—I would like to thank you for your careful critique of this thesis. Additionally, Prof. Dickhoff provided valuable guidance over the course of my Ph.D. as a member of my faculty mentoring committee, for which I am greatly appreciative.

So many opportunities over the course of this thesis would not have been possible without the work of the staff in the Department of Physics. I want to thank everyone in the main office for the work that you do to make the department run smoothly every day.

I also appreciated my friends and office-mates who made the Ph.D. an enjoyable experience. Particularly, I would like to thank Abe, Alex, Andrea, Anna, Graham, Jason, J. R., Liam, Lorenzo, Sam, Takuya, and Ziyuan. I am grateful that I had a chance to know you all and I appreciate the friendships made over the years.

In addition to the support that I received within the department, there are also a number of people external to Washington University deserve thanks. First, I want to acknowledge the great mentoring and instruction that I had at all levels of my academic career. This journey began thanks to the early encouragement from my high school physics teacher, Mrs. Harper, who made physics "phun." I am also grateful for guidance early in my career from Filomena Nunes, Amy Lovell, and Kendall Mahn.

Thanks to the support of the Stewardship Science Graduate Fellowship, I was able to carry out a research practicum at Los Alamos National Laboratory. I am grateful again to my hosts, Stefano and Joe, for mentoring me during my visit. I would like to thank the whole T-2 Division for their hospitality and, in particular, Emanuele, Ingo, Josh, Rahul, and Ronen for making me feel welcome and included while I was at the lab.

Of course, none of this would be possible without my family and friends. To my sister, Lauren, and my parents, Gale and Debby, thank you for your love and encouragement. To Jacob Boeck, Gurveer Deol, Jack Hamel, Brandon Lin, and Garrett Shafer, I am truly grateful for your consistent support and our friendship.

Finally, I am appreciative of the financial support that allowed me to carry out my research. I would like to acknowledge support by the US Department of Energy (DOE).

Washington University
St. Louis, MO, USA
August 2024

Garrett B. King

Competing Interests The author has no competing interests to declare that are relevant to the content of this manuscript.

Contents

Chapter 1
Introduction

Abstract In this chapter, I introduce the main topics of this thesis. I discuss ongoing and planned searches for beyond Standard Model physics, and the role that nuclear structure plays in interpreting these experiments. I also overview the recent developments in nuclear interactions and nuclear structure over the last decades, as well as the concept of quantum Monte Carlo methods. Finally, I lay out the structure of the thesis.

The Standard Model (SM) of particle physics is a highly successful theory that describes three out of the four presently known fundamental forces. While the SM accounts for a majority of known electromagnetic, strong, and weak processes, it cannot describe all observed physical phenomena. One clear example is the existence of neutrino oscillations [1, 2], a phenomenon that requires the neutrinos to have a non-zero mass. Within the SM, there is no mechanism to generate neutrino masses and, therefore, this cannot be the most fundamental description of nature. Additionally, the matter-antimatter imbalance in the universe and the nature of dark matter do not have adequate SM descriptions. Discovering the mechanisms that are responsible for these "beyond the SM" (BSM) phenomena will be a major focus of physics research in the years to come.

When attempting to identify the missing physics, there are several places where the community may look. One active area of research involves experimental searches for violations of the fundamental symmetries of the SM. This requires precise measurements of processes allowed in the SM so that small deviations from theoretical predictions can be detected. Nuclear β-decay—a process in which a neutron (proton) is converted into a proton (neutron), an electron (a positron), and an antineutrino (neutrino) in the final state—will receive renewed experimental attention. Deviations in predicted decay rates and decay spectra can provide evidence of BSM processes and help to pin down the Lorentz character of the operators responsible for mediating them [3–5].

Another class of experimental searches hunt for rare processes not currently allowed by the SM. For instance, measuring neutrinoless double β-decay ($0\nu\beta\beta$)—where two neutrons convert to two protons, two electrons, and no neutrinos in the final state, thus violating lepton number by two units—would provide a clear signature for the Majorana nature of the neutrino and necessarily involve new physics.

G. B. King, *Electroweak Structure and Reactions in Light Nuclei with Quantum Monte Carlo Methods*, Springer Theses, https://doi.org/10.1007/978-3-032-03191-4_1

One of the major difficulties of these searches is the need for nuclear matrix elements to interpret data because they can vary by factors of two to three depending on the method used to calculate them [6–8]. Thus, improvements to nuclear theory will serve to benefit these searches and make it possible to reliably extract new physics parameters.

Lastly, we may look toward the BSM phenomenon that experiments have already observed—namely, neutrino oscillations—to get a better understanding of the process. Planned and on-going long-baseline neutrino oscillation experiments [9–16]—aim to more precisely determine parameters that govern the neutrino flavor mixing and provide information related to neutrino masses. The oscillation probability depends on these parameters, as well as the length of the baseline and the incident neutrino energy. It is possible to fix the length of the baseline in the experiment; however, neutrinos are not produced in a mono-energetic beam. Rather, a flux of neutrino energies is created. This then necessitates the use of neutrino-nucleus cross sections to infer the energy of incident neutrinos required to determine oscillation parameters [17]. These cross sections, however, depend on the nuclear model and this remains a source of uncertainty in the experimental analysis.

All of the aforementioned efforts to search for new physics depend on theoretical determinations of SM processes involving nuclei in order to reliably and meaningfully interpret signatures of new physics. It is thus critical to have an accurate theoretical understanding of how the underlying dynamics of the protons and neutrons (collectively, the nucleons) inside of the nucleus impact electroweak processes. Given the importance of these endeavors not only to fundamental physics research, but also to the long-term goals of nuclear science [18], it is necessary to develop a robust theory of nuclei with well-understood model dependencies over a wide range of kinematics in order to disentangle signals of BSM physics from nuclear effects.

In this thesis, I will focus on a microscopic approach to modeling nuclei starting from the interaction of nucleonic degrees of freedom. This theory relies on three inputs: First, it needs a description of the interaction between nucleons in pairs, triples, and, in principle, larger groupings that is grounded in the underlying theory of quantum chromodynamics (QCD). Secondly, the approach requires an interaction between single nucleons, pairs of correlated nucleons, and clusters of correlated nucleons with external probes, such as electrons, photons, and neutrinos. Finally, it is necessary to develop computational methods to compute observable quantities and connect the theory with experiment. The reach of so-called "*ab initio*" nuclear theory—an umbrella under which the described microscopic procedure would fall—has expanded immensely in the last decade and includes several computational approaches to solve the nuclear many-body problem [19]. The techniques that I will describe in this thesis are in the class of quantum Monte Carlo (QMC) methods [20]—that is, stochastic approaches to solve the many-body Schrödinger Equation.

While the described microscopic approach depends on a high-quality description of the interaction between nucleons, these, of course, are not the most fundamental degrees of freedom for the strong force in the SM. Although QCD—which is described in terms of quark and gluon degrees of freedom—is presently the most fundamental theory of the strong interaction, it is only possible to describe nuclear structure and dynamics from these approaches for a select few light nuclei (presently, systems with mass number $A \leq 3$). Though tremendous progress has been made in describing nuclear structure [21, 22] and weak processes [23–27] with Lattice QCD computational techniques, the current best approach to directly study nuclei is a framework retaining interacting nucleons as the primary degrees of freedom.

Historically, the first effort to describe such an interaction was that of Hideki Yukawa who suggested that nucleon-nucleon (NN) force could be described in terms of the exchange of a particle that would come to be known as the pion, the lightest meson [28]. Though the one-pion exchange (OPE) picture gives rise to important features of the nucleon-nucleon (NN) interaction—such as its finite range, long-range attraction, and the presence of a tensor force—it is not adequate to describe all nuclear phenomena. Other sophisticated mechanisms are required to account for the complex structure of the NN interaction; for instance, interactions like the CD-Bonn potential include exchanges of heavier mesons, like the η, ρ, and ω, as well as that of a fictions scalar-isoscalar σ boson [29]. Phenomenological interactions, on the other hand, contain long-range pion exchange terms plus other operator structures that account for intermediate-range and short-range physics. In the latter two terms, parameters governing the strength of different contributions to the interactions may be fit to NN scattering data, the most notable example being the Argonne-v_{18} (AV18) potential that fit 4301 pp and np scattering data points with a χ^2/datum of 1.09 [30]. The existence of three-nucleon ($3N$) forces in addition to the NN interaction has been known since the 1930s [31] and their inclusion is a necessary component for nuclear interactions to reproduce the trinucleon binding energies [32]. In the phenomenological approach, this was typically achieved by supplementing the AV18 with three-nucleon forces—either the Urbana or Illinois models of the three-nucleon force [33, 34]. The phenomenological approaches also include one-body charge and current operators operators from the non-relativistic reduction of nucleon covariant currents along with two-body operators that arise to satisfy the continuity equation [35] as the basis to study electroweak phenomena.

Although phenomenological models achieved considerable success in describing nuclear systems and many-nucleon effects, chiral effective field theory (χEFT) has become increasingly popular in recent years. The χEFT approach is rooted in the fundamental symmetries of the strong interactions and provides a direct connection between many-nucleon interactions, electroweak currents and the underlying theory of QCD [36]. Further, this approach is based on a power expansion in a small parameter; namely, the ratio of the typical nucleon momentum ($\sim$100 MeV/c) to the typical scale of QCD ($\sim$1 GeV). The perturbative nature of χEFT allows, in principle, a systematic expansion of the nuclear interaction and electroweak currents. This, in turn, makes it possible to reliably estimate theoretical uncertainties and is an effort being pursued by the nuclear theory community [37–41] In this thesis, I will focus on the Norfolk two- and three-body local chiral interaction (NV2+3) [42–45] and associated electroweak currents [45–52].

Using QMC and the NV2+3, it is possible to validate the nuclear model and make predictions of observable quantities. In this thesis, I will discuss model validation related to the three classes of fundamental physics searches discussed in this introduction. First, I will present several studies of nuclear β-decay. This process is not only important to BSM physics experiments, but also to many-body theorists because the process has plenty of readily available experimental data. Therefore, reproducing measured β-decay rates and strengths was an important initial step for the NV2+3 model before proceeding to the first calculation of the ^{6}He β-decay spectrum to include the effect of many-body transition operators in the weak currents. This quantity is relevant to new physics searches looking from deviations from the SM in its measurement. While β-decays involve relatively low momentum transfers, $0\nu\beta\beta$ involves momentum transfers on the order of 100 MeV/c [53]. Thus, one should validate nuclear models on processes in this kinematic regime, as well. Fortuitously, muon captures on the nucleus involve a momentum transfer of the order of $m_\mu c \approx 105$ MeV/c (m_μ being the muon rest mass energy) and have readily available data. Therefore, I computed muon capture rates in light nuclei to validate the NV2+3 for those kinematics. Lastly, neutrino-nucleus scattering involves large momentum and energy transfers. The final set of results in this thesis focus on the magnetic structure of nuclei compared to experimental data. Validating on this process allowed for a test of the NV2+3 up to momentum transfers of 600 MeV/c, which will be a relevant kinematic regime for future neutrino oscillations experiments [15, 16]. Through these model validations, we have identified the strengths and shortcoming of the NV2+3 model over a wide range of kinematics. This will aid interpretations of experiments with kinematics for which the model is well-understood and accurate. Further, these studies have also indicated possible paths forward to improve future QMC calculations and the next generations of the χEFT potential.

This thesis is structured as follows: In Chap. 2, I will describe the salient features of the nuclear force and the derivation of potentials and current in χEFT. The specifics of the NV2+3 interaction are discussed in Sect. 2.2. In Chap. 3, I discuss the computation of nuclear matrix elements using QMC techniques. With the theoretical background covered, I present calculations of Gamow-Teller β-decays using QMC with the NV2+3 model in Chap. 4. Following this, in Chap. 5, I discuss my calculations of partial muon capture rates. Next, in Chap. 6, I review the validation of the electromagnetic current operator over a wide range of momentum transfer with calculations of magnetic moments and form factors. Finally, I conclude this work and provide an outlook in Chap. 7.

For ease of reading, these chapters will focus only on the physics necessary to understand the work carried out in this thesis, and thus I relegate more detailed information to several appendices. Namely, I provide detailed expressions of the NV2+3 potential and electroweak current operators in Appendix A. I present the derivation of the multipole decomposition of a generic charge and current operator, as well those of β-decay and muon capture rates, in Appendix B. Appendix C covers the details of the radiative corrections to β-decay to supplement the discussion of the ^{6}He β-decay spectrum in Sect. 4.3. Lastly, Appendix E provides the explicit multipole expressions of the magnetic form factors in light nuclei that were computed using QMC methods.

References

1. Y. Fukuda et al., Evidence for oscillation of atmospheric neutrinos. Phys. Rev. Lett. **81**, 1562–1567 (1998). https://doi.org/10.1103/PhysRevLett.81.1562
2. Q.R. Ahmad et al., Direct evidence for neutrino flavor transformation from neutral current interactions in the Sudbury Neutrino Observatory. Phys. Rev. Lett. **89**, 011301 (2002). https://doi.org/10.1103/PhysRevLett.89.011301
3. M. Gonzalez-Alonso, O. Naviliat-Cuncic, N. Severijns, New physics searches in nuclear and neutron β decay. Prog. Part. Nucl. Phys. **104**, 165–223 (2019). https://doi.org/10.1016/j.ppnp.2018.08.002
4. A. Falkowski, M. González-Alonso, O. Naviliat-Cuncic, Comprehensive analysis of beta decays within and beyond the Standard Model. JHEP **04**, 126 (2021). https://doi.org/10.1007/JHEP04(2021)126
5. M. Brodeur et al., Nuclear β decay as a probe for physics beyond the Standard Model. 1 (2023)
6. J. Engel, J. Menéndez, Status and Future of Nuclear Matrix Elements for Neutrinoless Double-Beta Decay: A Review. Rept. Prog. Phys. **80**(4), 046301 (2017). https://doi.org/10.1088/1361-6633/aa5bc5
7. V. Cirigliano et al., Towards precise and accurate calculations of neutrinoless double-beta decay. J. Phys. G **49**(12), 120502 (2022). https://doi.org/10.1088/1361-6471/aca03e
8. M. Agostini, G. Benato, J.A. Detwiler, J. Menéndez, F. Vissani, Toward the discovery of matter creation with neutrinoless $\beta\beta$ decay. Rev. Mod. Phys. **95**(2), 025002 (2023). https://doi.org/10.1103/RevModPhys.95.025002
9. K. Abe, R. Akutsu, A. Ali, C. Alt, C. Andreopoulos, L. Anthony, M. Antonova, S. Aoki, A. Ariga, Y. Asada, et al., Constraint on the matter-antimatter symmetry-violating phase in neutrino oscillations. *arXiv preprint* arXiv:1910.03887 (2019)
10. M.A. Acero, P. Adamson, L. Aliaga, T. Alion, V. Allakhverdian, S. Altakarli, N. Anfimov, A. Antoshkin, A. Aurisano, A. Back, C. Backhouse, M. Baird, N. Balashov, P. Baldi, B.A. Bambah, S. Bashar, K. Bays, S. Bending, R. Bernstein, V. Bhatnagar, B. Bhuyan, J. Bian, T. Blackburn, J. Blair, A.C. Booth, P. Bour, C. Bromberg, N. Buchanan, A. Butkevich, S. Calvez, M. Campbell, T.J. Carroll, E. Catano-Mur, A. Cedeno, S. Childress, B.C. Choudhary, B. Chowdhury, T.E. Coan, M. Colo, J. Cooper, L. Corwin, L. Cremonesi, G.S. Davies, P.F. Derwent, P. Ding, Z. Djurcic, D. Doyle, E.C. Dukes, H. Duyang, S. Edayath, R. Ehrlich, M. Elkins, G.J. Feldman, P. Filip, W. Flanagan, M.J. Frank, H.R. Gallagher, R. Gandrajula, F. Gao, S. Germani, A. Giri, R.A. Gomes, M.C. Goodman, V. Grichine, M. Groh, R. Group, B. Guo, A. Habig, F. Hakl, J. Hartnell, R. Hatcher, A. Hatzikoutelis, K. Heller, J. Hewes, A. Himmel, A. Holin, B. Howard, J. Huang, J. Hylen, F. Jediny, C. Johnson, M. Judah, I. Kakorin, D. Kalra, D.M. Kaplan, R. Keloth, O. Klimov, L.W. Koerner, L. Kolupaeva, S. Kotelnikov, I. Kourbanis, A. Kreymer, Ch. Kulenberg, A. Kumar, C.D. Kuruppu, V. Kus, T. Lackey, K. Lang, S. Lin, M. Lokajicek, J. Lozier, S. Luchuk, K. Maan, S. Magill, W.A. Mann, M.L. Marshak, M. Martinez-Casales, V. Matveev, D.P. Méndez, M.D. Messier, H. Meyer, T. Miao, W.H. Miller, S.R. Mishra, A. Mislivec, R. Mohanta, A. Moren, L. Mualem, M. Muether, S. Mufson, K. Mulder, R. Murphy, J. Musser, D. Naples, N. Nayak, J.K. Nelson, R. Nichol, G. Nikseresht, E. Niner, A. Norman, T. Nosek, A. Olshevskiy, T. Olson, J. Paley, R.B. Patterson, G. Pawloski, D. Pershey, O. Petrova, R. Petti, D.D. Phan, R.K. Plunkett, B. Potukuchi, C. Principato, F. Psihas, A. Radovic, V. Raj, R.A. Rameika, B. Rebel, P. Rojas, V. Ryabov, O. Samoylov, M.C. Sanchez, S. Sánchez Falero, I.S. Seong, P. Shanahan, A. Sheshukov, P. Singh, V. Singh, E. Smith, J. Smolik, P. Snopok, N. Solomey, E. Song, A. Sousa, K. Soustruznik, M. Strait, L. Suter, A. Sutton, R.L. Talaga, B. Tapia Oregui, P. Tas, R.B. Thayyullathil, J. Thomas, E. Tiras, D. Torbunov, J. Tripathi, A. Tsaris, Y. Torun, J. Urheim, P. Vahle, J. Vasel, L. Vinton, P. Vokac, T. Vrba, M. Wallbank, B. Wang, T.K. Warburton, M. Wetstein, M. While, D. Whittington, S.G. Wojcicki, J. Wolcott, N. Yadav, A. Yallappa Dombara, K. Yonehara, S. Yu, S. Zadorozhnyy, J. Zalesak, B. Zamorano, R. Zwaska, First measurement of neutrino oscillation parameters using neutrinos and antineutrinos by nova. Phys. Rev. Lett. **123**, 151803 (2019). https://doi.org/10.1103/PhysRevLett.123.151803

11. R. Acciarri, C. Adams, R. An, A. Aparicio, S. Aponte, J. Asaadi, M. Auger, N. Ayoub, L. Bagby, B. Baller et al., Design and construction of the microboone detector. J. Instrum. **12**(02), P02017 (2017)
12. L. Aliaga, L. Bagby, B. Baldin, A. Baumbaugh, A. Bodek, R. Bradford, W.K. Brooks, D. Boehnlein, S. Boyd, H. Budd et al., Design, calibration, and performance of the minerva detector. Nucl. Instrum. Methods Phys. Res., Sect. A **743**, 130–159 (2014)
13. S. Seon-Hee, Physics potentials of the hyper-kamiokande second detector in korea. *arXiv preprint* arXiv:1811.06682 (2018)
14. B. Abi, R. Acciarri, M.A. Acero, G. Adamov, D. Adams, M. Adinolfi, Z. Ahmad, J. Ahmed, T. Alion, S. Alonso Monsalve, et al., Deep underground neutrino experiment (dune), far detector technical design report, volume ii dune physics. *arXiv preprint* arXiv:2002.03005 (2020)
15. A. Abed Abud et al., Snowmass Neutrino Frontier: DUNE Physics Summary. 3 (2022)
16. J. Bian et al., Hyper-Kamiokande Experiment: A Snowmass White Paper. In Snowmass **2021**, 3 (2022)
17. L. Alvarez-Ruso et al., NuSTEC White Paper: Status and challenges of neutrino-nucleus scattering. Prog. Part. Nucl. Phys. **100**, 1–68 (2018). https://doi.org/10.1016/j.ppnp.2018.01.006
18. Nuclear Science Advisory Committee. A new era of discovery: The 2023 long range plan for nuclear science. https://science.osti.gov/np/nsac/Reports
19. H. Hergert, A Guided Tour of *ab initio* Nuclear Many-Body Theory. Front. Phys. **8**, 379 (2020). https://doi.org/10.3389/fphy.2020.00379
20. J. Carlson, S. Gandolfi, F. Pederiva, Steven C. Pieper, R. Schiavilla, K. E. Schmidt, and R. B. Wiringa. Quantum Monte Carlo methods for nuclear physics. Rev. Mod. Phys. **87**, 1067 (2015). https://doi.org/10.1103/RevModPhys.87.1067
21. P.E. Shanahan, Chiral Effective Theory Methods and their Application to the Structure of Hadrons from Lattice QCD. J. Phys. G **43**(12), 124001 (2016). https://doi.org/10.1088/0954-3899/43/12/124001
22. W. Detmold and P. E. Shanahan. Few-nucleon matrix elements in pionless effective field theory in a finite volume. 2 2021
23. M.J. Savage, P.E. Shanahan, B.C. Tiburzi, M.L. Wagman, F. Winter, S.R. Beane, E. Chang, Z. Davoudi, W. Detmold, K. Orginos, Proton-Proton Fusion and Tritium β Decay from Lattice Quantum Chromodynamics. Phys. Rev. Lett. **119**(6), 062002 (2017). https://doi.org/10.1103/PhysRevLett.119.062002
24. B.C. Tiburzi, M.L. Wagman, F. Winter, E. Chang, Z. Davoudi, W. Detmold, K. Orginos, M.J. Savage, P.E. Shanahan, Double-β Decay Matrix Elements from Lattice Quantum Chromodynamics. Phys. Rev. D **96**(5), 054505 (2017). https://doi.org/10.1103/PhysRevD.96.054505
25. Z. Davoudi, W. Detmold, K. Orginos, A. Parreño, M.J. Savage, P. Shanahan, M.L. Wagman, Nuclear matrix elements from lattice QCD for electroweak and beyond-Standard-Model processes. 8 2020. https://dx.doi.org/10.1016/j.physrep.2020.10.004
26. V. Cirigliano, W. Detmold, A. Nicholson, P. Shanahan, Lattice QCD Inputs for Nuclear Double Beta Decay. 3 2020. https://doi.org/10.1016/j.ppnp.2020.103771
27. A. Parreño, P.E. Shanahan, M.L. Wagman, F. Winter, E. Chang, W. Detmold, M. Illa, The axial charge of the triton from lattice QCD. 2 (2021)
28. H. Yukawa, On the Interaction of Elementary Particles I. Proc. Phys. Math. Soc. Jap. **17**, 48–57 (1935). https://doi.org/10.1143/PTPS.1.1
29. R. Machleidt, High-precision, charge-dependent bonn nucleon-nucleon potential. Phys. Rev. C **63**, 024001 (2001). https://doi.org/10.1103/PhysRevC.63.024001
30. R.B. Wiringa, V.G.J. Stoks, R. Schiavilla, An Accurate nucleon-nucleon potential with charge independence breaking. Phys. Rev. C **51**, 38–51 (1995). https://doi.org/10.1103/PhysRevC.51.38
31. H. Primakoff, T. Holstein, Many-body interactions in atomic and nuclear systems. Phys. Rev. **55**, 1218–1234 (1939). https://doi.org/10.1103/PhysRev.55.1218

32. J.L. Friar, B.F. Gibson, G.L. Payne, Recent progress in understanding trinucleon properties. Ann. Rev. Nucl. Part. Sci. **34**, 403–433 (1984). https://doi.org/10.1146/annurev.ns.34.120184.002155
33. R.B. Wiringa, Interplay between two-and three-body interaction in light nuclei and nuclear matter. Nuc. Phys. A **401**(1), 86–92 (1983)
34. S.C. Pieper, V.R. Pandharipande, R.B. Wiringa, J. Carlson, Realistic models of pion exchange three nucleon interactions. Phys. Rev. C **64**, 014001 (2001). https://doi.org/10.1103/PhysRevC.64.014001
35. D. Olof Riska, EXCHANGE CURRENTS. Phys. Rept. **181**, 207 (1989). https://doi.org/10.1016/0370-1573(89)90129-4
36. F. Gross et al. 50 Years of Quantum Chromodynamics. 12 (2022)
37. R.J. Furnstahl, N. Klco, D.R. Phillips, S. Wesolowski, Quantifying truncation errors in effective field theory. Phys. Rev. C **92**(2), 024005 (2015). https://doi.org/10.1103/PhysRevC.92.024005
38. A. Ekström, G. Hagen, Global sensitivity analysis of bulk properties of an atomic nucleus. Phys. Rev. Lett. **123**(25), 252501 (2019). https://doi.org/10.1103/PhysRevLett.123.252501
39. D.R. Phillips et al., Get on the BAND Wagon: A Bayesian Framework for Quantifying Model Uncertainties in Nuclear Dynamics. J. Phys. G **48**(7), 072001 (2021). https://doi.org/10.1088/1361-6471/abf1df
40. J.A. Melendez, R.J. Furnstahl, D.R. Phillips, M.T. Pratola, S. Wesolowski, Quantifying Correlated Truncation Errors in Effective Field Theory. Phys. Rev. C **100**(4), 044001 (2019). https://doi.org/10.1103/PhysRevC.100.044001
41. S. Wesolowski, I. Svensson, A. Ekström, C. Forssén, R.J. Furnstahl, J.A. Melendez, D.R. Phillips, Rigorous constraints on three-nucleon forces in chiral effective field theory from fast and accurate calculations of few-body observables. Phys. Rev. C **104**(6), 064001 (2021). https://doi.org/10.1103/PhysRevC.104.064001
42. M. Piarulli, L. Girlanda, R. Schiavilla, R. Navarro Pérez, J. E. Amaro, and E. Ruiz Arriola. Minimally nonlocal nucleon-nucleon potentials with chiral two-pion exchange including Δ resonances. Phys. Rev. **C91**(2), 024003 (2015). https://doi.org/10.1103/PhysRevC.91.024003
43. M. Piarulli, L. Girlanda, R. Schiavilla, A. Kievsky, A. Lovato, L.E. Marcucci, S.C. Pieper, M. Viviani, R.B. Wiringa, Local chiral potentials with Δ-intermediate states and the structure of light nuclei. Phys. Rev. C **94**(5), 054007 (2016). https://doi.org/10.1103/PhysRevC.94.054007
44. M. Piarulli et al., Light-nuclei spectra from chiral dynamics. Phys. Rev. Lett. **120**(5), 052503 (2018). https://doi.org/10.1103/PhysRevLett.120.052503
45. A. Baroni et al., Local chiral interactions, the tritium Gamow-Teller matrix element, and the three-nucleon contact term. Phys. Rev. C **98**(4), 044003 (2018). https://doi.org/10.1103/PhysRevC.98.044003
46. S. Pastore, R. Schiavilla, J.L. Goity, Electromagnetic two-body currents of one- and two-pion range. Phys. Rev. C **78**, 064002 (2008). https://doi.org/10.1103/PhysRevC.78.064002
47. S. Pastore, L. Girlanda, R. Schiavilla, M. Viviani, R.B. Wiringa, Electromagnetic Currents and Magnetic Moments in (chi)EFT. Phys. Rev. C **80**, 034004 (2009). https://doi.org/10.1103/PhysRevC.80.034004
48. S. Pastore, L. Girlanda, R. Schiavilla, M. Viviani, The two-nucleon electromagnetic charge operator in chiral effective field theory (χEFT) up to one loop. Phys. Rev. C **84**, 024001 (2011). https://doi.org/10.1103/PhysRevC.84.024001
49. M. Piarulli, L. Girlanda, L.E. Marcucci, S. Pastore, R. Schiavilla, M. Viviani, Electromagnetic structure of A = 2 and 3 nuclei in chiral effective field theory. Phys. Rev. C **87**(1), 014006 (2013). https://doi.org/10.1103/PhysRevC.87.014006
50. A. Baroni, L. Girlanda, S. Pastore, R. Schiavilla, M. Viviani, Nuclear Axial Currents in Chiral Effective Field Theory. Phys. Rev. C **93**(1), 015501 (2016). https://doi.org/10.1103/PhysRevC.93.049902, https://doi.org/10.1103/PhysRevC.93.015501,10.1103/PhysRevC.95.059901. [Erratum: Phys. Rev. C 95, no.5,059901(2017)]
51. A. Baroni, L. Girlanda, A. Kievsky, L.E. Marcucci, R. Schiavilla, M. Viviani, Tritium β-decay in chiral effective field theory. Phys. Rev. C **94**(2), 024003 (2016). https://doi.org/10.1103/PhysRevC.94.024003, https://doi.org/10.1103/PhysRevC.95.059902. [Erratum: Phys. Rev. C 95, no.5,059902(2017)]

52. R. Schiavilla, A. Baroni, S. Pastore, M. Piarulli, L. Girlanda, A. Kievsky, A. Lovato, L. E. Marcucci, Steven C. Pieper, M. Viviani, and R. B. Wiringa. Local chiral interactions and magnetic structure of few-nucleon systems. Phys. Rev. C **99**, 034005 (2019). https://doi.org/10.1103/PhysRevC.99.034005
53. S. Pastore, J. Carlson, V. Cirigliano, W. Dekens, E. Mereghetti, R.B. Wiringa, Neutrinoless double-β decay matrix elements in light nuclei. Phys. Rev. C **97**(1), 014606 (2018). https://doi.org/10.1103/PhysRevC.97.014606

Chapter 2
Nuclear Model from Chiral Effective Field Theory

Abstract In this chapter, I will detail the chiral effective field theory (χEFT) approach to derive nuclear interactions and electroweak currents. This approach—rooted in the symmetries of the QCD Lagrangian—has become popular in the nuclear theory community. There are a number of recent reviews of χEFT [1–3] and it is beyond the scope of this thesis to discuss the formalism in detail; therefore, I will review only the salient points of χEFT potentials and currents with a particular emphasis on the properties of the fully local Norfolk two- and three-body (NV2+3) potentials [4–7] and consistent electroweak transition operators [7–14] that I use throughout this thesis.

2.1 Nuclear Hamiltonian

The many-body Hamiltonian for the nuclear system is typically modelled as,

$$H = \sum_i K_i + \sum_{i<j} v_{ij} + \sum_{i<j<k} V_{ijk} + \dots , \tag{2.1}$$

where K_i is the single nucleon non-relativistic kinetic energy and v_{ij} and V_{ijk} are two- and three-nucleon potentials, respectively. Similarly, the description of the interactions of the nucleus with external electroweak probes requires both many-nucleon charge and current operators, ρ and $\mathbf{j}$, respectively. Schematically, they may be written as,

$$\begin{aligned} \rho &= \sum_i \rho_i(\mathbf{q}) + \sum_{i<j} \rho_{ij}(\mathbf{q}) + \dots , \\ \mathbf{j} &= \sum_i \mathbf{j}_i(\mathbf{q}) + \sum_{i<j} \mathbf{j}_{ij}(\mathbf{q}) + \dots , \end{aligned} \tag{2.2}$$

where $\mathbf{q}$ is the momentum transferred to the nucleus and the interaction with the external field may be decomposed into interactions with a single nucleon and correlated pairs of nucleons.

G. B. King, *Electroweak Structure and Reactions in Light Nuclei with Quantum Monte Carlo Methods*, Springer Theses, https://doi.org/10.1007/978-3-032-03191-4_2

While the focus of this section will primarily be the specific two- and three-nucleon interactions used in this work—the NV2+3 models—it is instructive to discuss the phenomenological understanding of the nuclear force. Prior to χEFT, the typical approach to building two-nucleon (*NN*) interactions began by considering all of the symmetries obeyed by this system and constructing a general set of operators upon which the force could depend [15]. In the absence of an external field applied to the system, the NN interaction is invariant under spatial translations, therefore restricting the coordinate dependence to the relative distance $r = |\mathbf{r}_i - \mathbf{r}_j|$ of the two particles denoted as i and j. For non-relativistic *NN* interactions, Galilean invariance restricts any velocity dependence to be relative, as well. Thus, the potential may only depend on the relative momentum $\mathbf{p} = \mathbf{p}_i - \mathbf{p}_j$ between the two particles. Invariance under rotations allows for the description of the two nucleon system in terms of total angular momentum J and its projection M. Because protons (p) and neutrons (n) are spin $1/2$ particles and, thus, fermions, there is a restriction on the allowed values of JM for the pp, nn, and np systems. For instance, using the standard $^{2S+1}L_J$ spectroscopic notation—where S and L denote the total spin and orbital angular momentum quantum numbers—the np system may couple to the $^1S_0$ and $^3S_1$ channels when $L = 0$ while the pp and nn systems are restricted only to $^1S_0$. Parity is conserved by the strong force and this symmetry must also be obeyed by the *NN* interaction. Finally, the force satisfies time-reversal invariance.

Because of the relatively small mass splitting between the protons and neutrons, as well as the symmetry in the np and pp scattering when the Coulomb contribution is removed [16], it is natural to propose that protons and neutrons are two different projections of a spin-like symmetry; namely, isospin [17]. Pauli matrices $\boldsymbol{\sigma}_i$ and $\boldsymbol{\tau}_i$ represent, respectively, the spin and isospin of nucleon i. Assuming charge-independence (CI) of the *NN* interaction implies that the potential may only depend on isoscalar combinations of $\boldsymbol{\tau}_i$ and $\boldsymbol{\tau}_j$ [15]. Charge-symmetry (CS), instead, implies that the potential is the same when all protons are swapped for neutrons and vice versa. Because these are approximate symmetries, there may be small contributions that break them.

Another important and long-established feature of the nuclear interaction is the need for a tensor force. This is most clearly illustrated by the presence of a non-vanishing quadrupole moment in the deuteron ground state [18]. As such, additional non-central terms are needed in the potential. Considering, for now, only static terms, such a dependence could be generated by the operator $\boldsymbol{\sigma}_i \cdot \hat{\mathbf{r}}$, where $\mathbf{r}$ is the internucleon distance. The dependence on this term cannot be linear because of the aforementioned parity and time-reversal symmetries. One thus introduces the only allowed combination of two such terms which is given in its standard form as the tensor operator,

$$S_{ij}(\hat{\mathbf{r}}) = 3(\boldsymbol{\sigma}_i \cdot \hat{\mathbf{r}})(\boldsymbol{\sigma}_j \cdot \hat{\mathbf{r}}) - \boldsymbol{\sigma}_i \cdot \boldsymbol{\sigma}_j \,, \tag{2.3}$$

which allows the admixture of orbital states with $\Delta L = 2$ having the same total angular momentum and parity [19]. This, for instance, gives the admixture of $^3S_1$

and $^3D_1$ waves present in the deuteron ground state. The most general, velocity-independent potential obeying the relevant symmetries and containing the tensor force is then schematically given by,

$$v_{ij} = \sum_{l=1}^{6} v_{ij}^{l}(r)\mathcal{O}_{ij}^{p}\,, \tag{2.4}$$

where $v_{ij}^{l}(r)$ are radial functions, l is an index labelling the different contributions, and the operators are defined as,

$$\mathcal{O}_{ij}^{l=1,\ldots,6} = [1, \boldsymbol{\sigma}_i \cdot \boldsymbol{\sigma}_j, S_{ij}(\hat{\mathbf{r}})] \otimes [1, \boldsymbol{\tau}_i \cdot \boldsymbol{\tau}_j]\,. \tag{2.5}$$

This operator structure includes terms arising from one-pion exchange, which is well-known as the source of the long-range attraction of the *NN* interaction [20]. Other important features of the potential are the intermediate range attraction from two-pion exchange (TPE) and heavier meson-exchange terms, a velocity dependence needed to describe *NN* scattering data in higher partial waves [21], and a repulsive core [15]. In addition, the underbinding of the trinucleons using only *NN* potentials necessitates three-body forces [22]. The traditional phenomenological approach augments the nuclear interaction by adding short and intermediate range terms in the *NN* sector and additional contributions from three-body forces. While phenomenological models like the Argonne potentials [21] had tremendous success in describing nuclear systems, these models are not systematically improvable and cannot be easily related to quantum chromodynamics (QCD), the fundamental theory underlying the nuclear interaction. The χEFT approach provides a means of systematically including terms in the *NN* interaction that is rooted in the underlying theory of QCD. Further, one can consistently derive electroweak transition operators that are linked to the *NN* interaction.

2.2 The Norfolk Model

The procedure for deriving the *NN* interaction from χEFT was pioneered by Weinberg in the 1990s [23]. This consists of writing down the most general Lagrangian consistent with the underlying symmetries of low-energy QCD for the chosen degrees of freedom. In the low-energy limit, the relevant degrees of freedom are the pion, with mass $m_\pi \approx 140\,\text{MeV}/c^2$, the nucleons with typical momentum $Q \approx m_\pi$, and the Δ-isobars, with mass splitting relative to the nucleons $m_{N\Delta} \approx 2m_\pi$. Heavier meson degrees of freedom occur with masses at the scale $\Lambda_\chi \approx 1$ GeV. This natural separation of scales implies that it is possible to integrate out the heavy meson degrees of freedom in low-energy calculations; Thus, the low-energy Lagrangian contains couplings involving nucleons, pions, and Δ-isobars and the physics of the heavy degrees of freedom are subsumed in coefficients governing the strength of operators that are known as "low-energy constants (LECs)."

Great effort has gone into deriving chiral Lagrangians and I direct interested readers to the details of Refs. [24–29]. For the purposes of the review, I simply note that, schematically, the chiral Lagrangian containing nucleon, pion, and Δ-isobar degrees of freedom admits a power expansion in Q/Λ_χ, given by

$$\mathcal{L} = \mathcal{L}_0 + \sum_{\nu=0}^{\infty} \mathcal{L}^{(\nu)}, \tag{2.6}$$

where $\mathcal{L}_0$ describes the dynamics of free pions, nucleons, and Δ-isobars and $\mathcal{L}^{(\nu)}$ indicates an interaction term of order ν in the power expansion. This then implies that one can derive a potential of the form,

$$v = \sum_{\nu=0}^{\infty} v^{(\nu)}. \tag{2.7}$$

At LO (Q^0), the potential consists of static OPE and contact interactions. The NLO term vanishes and the first subleading contributions is at N2LO (Q^2) from two-pion exchange (TPE) and contact (CT) interactions. The N3LO (Q^3) potential contains additional long-range and CT contributions [1, 2].

The first generation Norfolk two-nucleon (NV2) interaction [4, 5] was developed using the scheme detailed above. Together with the Norfolk three-nucleon (NV3) interaction [6, 7], this makes the fully local chiral NV2+3 interaction. Schematically, the NV2 interaction can be broken down into three parts,

$$v_{ij} = v_{ij}^L + v_{ij}^S + v_{ij}^{\mathrm{EM}}, \tag{2.8}$$

where v_{ij}^L is a long-range term containing OPE and TPE to order N2LO, v_{ij}^S is a short-range term containing CT interactions to order N2LO plus phenomenologically included CT terms from N3LO, and v_{ij}^{EM} is an electromagnetic term that has the same operator structure as that of the Argonne potentials [21]. Appendix A contains the explicit expressions of all radial dependencies in Sect. A.1 and diagrams that schematically represent the long-range strong interaction terms in the NV2 potential are shown in Fig. 2.1. The short-range CT terms involve nucleons interacting locally at a point with no intermediate pion or Δ-isobar propagators.

In quantum Monte Carlo (QMC) methods, it is much more efficacious to use a local interaction as compared to a non-local one [30]; however, when deriving the potential in χEFT, non-localities arise for two reasons: (i) the specific choice of regulator in momentum space and (ii) dependencies on the center of mass momentum transfer $\mathbf{K} = \frac{1}{2}(\mathbf{p} + \mathbf{p}')$ in contact interactions, where $\mathbf{p}$ and $\mathbf{p}'$ are the initial and final relative momentum, respectively. Using Fierz transformations to convert dependencies on $\mathbf{K}$ to ones on $\mathbf{k}$ and local regulators in coordinate space, the authors of Ref. [4] built

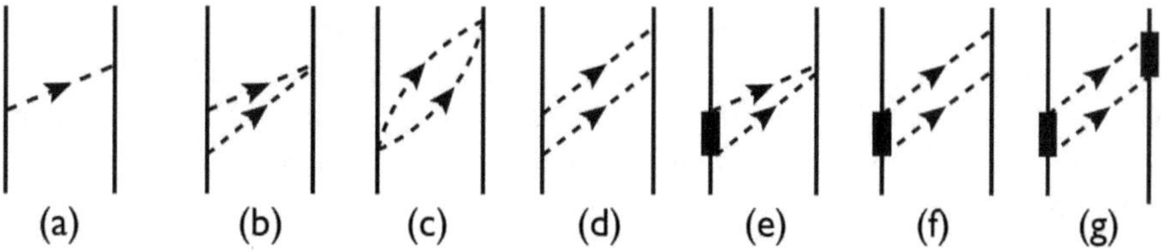

Fig. 2.1 Feynman diagrams representing different long-range contributions to the *NN* potential in χEFT. Solid lines represent nucleons, dashed lines represent pions, and thick solid lines represent intermediate Δ-isobars. Panel **a** represents the long-range OPE at LO (Q^0) in the potential. Beginning at N2LO (Q^2), there are contributions from TPE involving no (panels **b** through **d**), one (panels **e** and **f**), and two (panel **g**) intermediate Δ-isobars. Figure adapted with permission from Ref. [4]

a chiral interaction that is as minimally non-local as possible. In this scheme, it is possible to write an *NN* interaction that is fully local up to N2LO; however, at N3LO, non-locality is unavoidable [5]. The fully local picture was derived from the minimally non-local interaction by constraining the LECs governing the non-local terms at N3LO to vanish [5]. The end result is only including the local CT terms at N3LO necessary to describe *NN* scattering data while neglecting the long-range and non-local CT contribution at that order.

The resultant long-range potential has the form [5]

$$v_{ij}^L = \left[\sum_{l=1}^{6} v_L^l(r) O_{ij}^l \right] + v_L^{\sigma T}(r) O_{ij}^{\sigma T} + v_L^{tT}(r) O_{ij}^{tT} , \tag{2.9}$$

where $O_{ij}^{l=1,\ldots,6}$ are those defined in Eq. 2.4, $O_{ij}^{\sigma T} = \boldsymbol{\sigma}_i \cdot \boldsymbol{\sigma}_j T_{ij}$, $O_{ij}^{tT} = S_{ij} T_{ij}$, and $T_{ij} = 3\tau_{i,z}\tau_{j,z} - \boldsymbol{\tau}_i \cdot \boldsymbol{\tau}_j$. The first six terms are the charge independent central, spin, and tensor components with and without $\boldsymbol{\tau}_i \cdot \boldsymbol{\tau}_j$ dependence, while the latter terms are CI breaking central and tensor components arising due to the difference in the charged and neutral pion masses in OPE. The radial functions $v_L^l(r)$, described in detail in Appendix A of Ref. [4] and recapitulated in Sect. A.1 of this thesis, depend on the nucleon axial coupling g_A, the N–to–Δ axial coupling h_A, the neutral and charged pion masses m_{π_0} and m_{π_+}, and the pion decay amplitude F_π. To regulate singularities in the $v_L^l(r)$ that go like $1/r^n$ up to $n = 6$, they must be multiplied by a regulator function,

$$C_{R_L}(r) = 1 - \frac{1}{(r/R_L)^6 e^{(r-R_L)/a_L} + 1} , \tag{2.10}$$

with $a_L = R_L/2$. The short-range potential has the structure

$$v_{12}^S = \sum_{l=1}^{16} v_S^l(r) O_{ij}^l , \tag{2.11}$$

where $O_{ij}^{7,\ldots,11} = \mathbf{L}\cdot\mathbf{S},\ \mathbf{L}\cdot\mathbf{S}\boldsymbol{\tau}_i\cdot\boldsymbol{\tau}_j,\ (\mathbf{L}\cdot\mathbf{S})^2,\ \mathbf{L}^2,\ \mathbf{L}^2\boldsymbol{\sigma}_i\cdot\boldsymbol{\sigma}_j$, and $O_{ij}^{12,\ldots,16} = T_{ij},\ \tau_{i,z}+\tau_{j,z},\ \boldsymbol{\sigma}_i\cdot\boldsymbol{\sigma}_j T_{ij},\ S_{ij}T_{ij}$, and $\mathbf{L}\cdot\mathbf{S}T_{ij}$ for relative angular momentum $\mathbf{L} = \mathbf{r}\times\mathbf{p} = -i\mathbf{r}\times\nabla$ and total spin $\mathbf{S} = (\boldsymbol{\sigma}_i + \boldsymbol{\sigma}_j)/2$ of the *NN* system. Operators $O_{ij}^{l\geq 12}$ are CI and CS breaking contributions to the potential. The short-range regulator has the form,

$$C_{R_s} = \frac{1}{\pi^{3/2}R_S^3}e^{-(r/R_S)^2}\,, \tag{2.12}$$

which amounts to smearing out the delta functions in the CT term. In total, the short-range radial functions contain 26 unknown LECs that have been fit to low-energy *NN* cross-sections, polarization observables, and normalization data [5] from the Granada database [31–33].

Another feature of the χEFT approach is that one may obtain the $3N$ force in the same framework as the *NN* potential. Many-body forces appear in a hierarchical way in the chiral expansion, with the first $3N$ force being subleading relative to the LO term in the *NN* interaction, then the first four-nucleon term suppressed relative to the leading $3N$ term, and so on. The N2LO expression of the $3N$ force was derived by van Kolck et al. [24] and Epelbaum et al. [27]. In the Δ-full picture, the diagram involving the excitation of an intermediate Δ—corresponding to the Fujita-Miyazawa (FM) term [34]—is promoted to NLO in the chiral expansion [2].

Including the NV3 interaction introduces the FM term plus one TPE and two CT interaction diagrams to the potential, as seen in Fig. 2.2. The long-range diagrams again contain the known LECs from the long-range part of the NV2. The LECs in the short-range part of the interaction—c_D and c_E—are fit either to the triton ground state energy and n-deuteron (nd) doublet scattering length [6] or to the trition ground state energy and Gamow-Teller matrix element for β-decay [7]. This latter procedure is made possible due to a connection between the LEC z_0 (see Eq. A.106) governing the strength of the two-body axial contact operator at N3LO and the LEC c_D governing the contact coupling of two nucleons with a pion emitted in a relative p-wave [35–37].

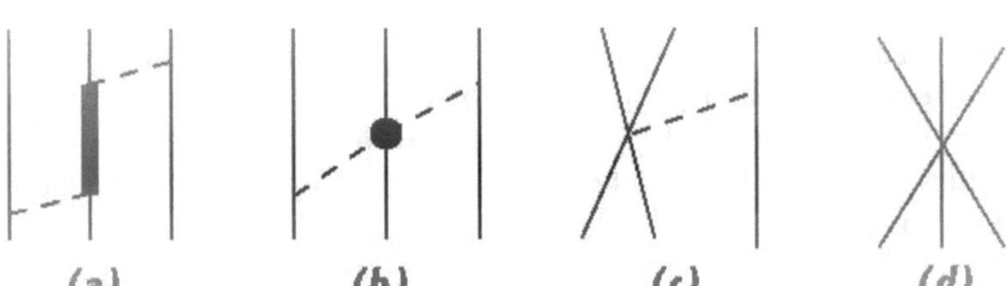

Fig. 2.2 Feynman diagrams representing different contributions to the $3N$ potential in χEFT. Solid lines represent nucleons, dashed lines represent pions, and thick solid lines represent intermediate Δ-isobars. Panel **a** represents the FW term at NLO (Q^1) in the chiral expansion, while the remaining diagrams in **b** through **d** comprise the N2LO (Q^2) contribution. Panel **b** depends on LECs from the πN Lagrangian, while panels **c** and **d** contain the unknown LECs c_D and c_E, respectively. Figure reproduced with permission from Ref. [6]

The different fitting procedures for the NV2+3 interaction lead to eight classes of the model. There are two classes (I and II) of the NV2 that differ only in the range of energy over which they are fitted—class I up to 125 MeV, and class II up to 200 MeV. For each class, two combinations of short- and long-range regulators were chosen. Two models correspond to "soft" cutoffs $(R_S, R_L) = (0.8, 1.2)$ fm (models NV2-Ia and NV2-IIa) and two more to "hard" cutoffs $(R_S, R_L) = (0.7, 1.0)$ fm (models NV2-Ib and NV2-IIb). Class I (II) fits about 2700 (3700) data points with a χ^2/datum $\lesssim 1.1$ ($\lesssim 1.4$) [4, 5]. If $3N$ LECs were fit with strong interaction data only (the trinucleons energies and the nd doublet scattering length), then no additional annotations are added to the model class. Models that fit strong and weak data simultaneously (the trinucleon energies and the tritium Gamow-Teller β-decay matrix element) are denoted with a star ($\star$).

2.3 Electroweak Currents

In addition to studying the interaction amongst nucleons, χEFT has been applied to the study of nuclear electroweak currents with the earliest efforts being those of Park, Min, and Rho in the 1990s [38, 39]. More recently, electroweak currents consistent with the modern NV2+3 interaction have been derived in Refs. [7–14] using time-ordered perturbation theory. This procedure involves computing the "transition" or "T-" matrix from the interaction Hamiltonian implied by the chiral Lagrangian describing the interactions between pions, nucleons, and Δ-isobars among themselves and with an external electroweak field $A^\mu = (A^0, \mathbf{A})$ having vector or axial(-vector) character under parity transformations [40, 41]. The T-matrix is given by,

$$\langle NN'|T|NN\rangle = \langle NN'|H_I \sum_{n=1}^{\infty} \left(\frac{1}{E_i - H_0 + i\eta} H_I \right)^{n-1} |NN\rangle \,, \tag{2.13}$$

where H_0 is the free Hamiltonian in the absence of interactions and the states $|NN\rangle$ and $|NN'\rangle$ are two nucleons states with energy E_i and $E_f = E_i$, respectively. The T-matrix expansion starts at order Q^{-3} due to the disconnected impulse-approximation diagrams at LO describing the coupling of a single nucleon to the external field, denoted in Fig. 2.3 with "1b(NR)." Schematically, it may be written as,

$$\mathcal{T}_e = \sum_{\nu=-3}^{\infty} \mathcal{T}_e^{(\nu)} \,. \tag{2.14}$$

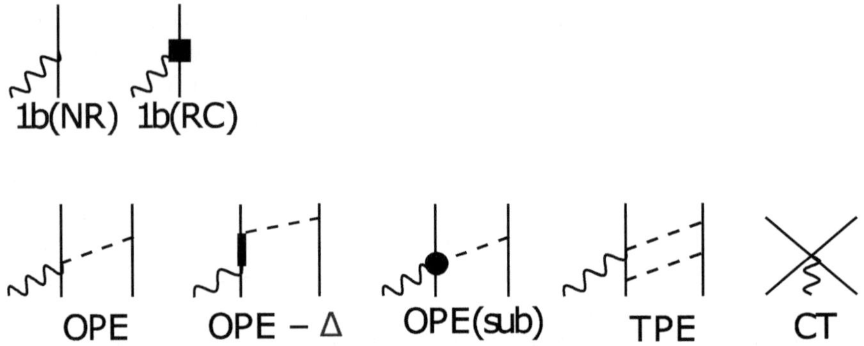

Fig. 2.3 Schematic representation of the electroweak currents in χEFT used in this work. Solid lines indicate nucleons, dashed lines indicate pions, thick solid lines indicate intermediate Δ-isobars, and wavy lines indicate an external vector or axial field. The square denotes a higher-order correction arising from the non-relativistic reduction of the covariant one-body current operator while the circle represents a subleading πNNA coupling in the χEFT Lagrangian. Figure reproduced with permission from Ref. [42]

One may solve the power expansion of the charge ρ and current $\mathbf{j}$ by defining $v_e = A^0\rho - \mathbf{A}\cdot\mathbf{j}$ with ρ and requiring that the on-energy-shell Lippmann-Schwinger equation is satisfied,

$$\mathcal{T} = (v + v_e) + (v + v_e)G_0(v + v_e) + (v + v_e)G_0(v + v_e)G_0(v + v_e) + \dots , \tag{2.15}$$

where v is the nuclear potential of Eq. 2.7. The charge and current are then extracted by an order-by-order matching to the T-matrix [10].

Along with the LO one-body operator consistent with the non-relativistic reduction of the nucleon covariant current, there is a one-body relativistic correction, denoted as "1b(RC)" in Fig. 2.3, appearing at N2LO. Additional contributions arise due to the interaction of two correlated nucleons with the external field. These two-body current operators may be understood intuitively as arising due to the exchange of quantum numbers induced by the nuclear force. This then leads to the movement of electromagnetic or weak charge in the nucleus not associated with the nucleon currents, thus necessitating higher-order corrections beyond the one-body picture. To the order used in this work, the two-body terms involve OPE diagrams, OPE with the excitation of an intermediate Δ-isobar (Δ-OPE), two pion loop diagrams, and CT diagrams. Because of a discrepancy in different approaches to obtain the N4LO axial charge and current [13], they are omitted in this work and all of the following calculations involving axial operators will be based on the tree-level contributions only. Figure 2.3 schematically shows the diagrams corresponding to different contributions in the charge and current operators while Table 2.1 gives an order-by-order breakdown of the axial and vector charge and current topologies.

To make connection to the physical processes which can be studied, I note that the Lorentz character of the electromagnetic current operator is vector, while the

Table 2.1 Order-by-order breakdown of the topologies contributing to the NV2+3 vector and axial charge and current operators [7, 10, 13, 14, 43]. The symbol "×" indicates a term that has been omitted in this work. The diagrams are labeled according to Fig. 2.3

Order	ρ_V	$\mathbf{j}_V$	ρ_A	$\mathbf{j}_A$
Q^{-3}	1b(NR)	–	–	1b(NR)
Q^{-2}	–	1b(NR)	1b(NR)	–
Q^{-1}	1b(RC)	OPE	OPE	1b(RC), Δ-OPE
Q^{0}	OPE	1b(RC), Δ-OPE	1b(RC)	OPE, CT
Q^{1}	LOOP	OPE, LOOP, CT	×	×

weak current is a vector-minus-axial operator. Because of the conserved vector current hypothesis, the operator structure of the weak vector current is identical to the electromagnetic vector current, differing only by a rescaling of coupling constants. Thus, taking the appropriate combinations of the operators allows one to study electromagnetic and weak processes.

Finally, I remark that the axial and vector charge and current operators, in principle, depend on LECs. In some cases, these come directly from the $NV2$; however, that is not always the case. For instance, as previously mentioned in the preceding section, the single unknown weak current LEC may be related to an unknown LEC of the $3N$ force. Therefore, it may either be taken from a fit of strong interaction data or included in the fit of the $3N$ term as a way to constrain both the current and the potential simultaneously. Instead, in the vector charge operator, there are OPE and CT contributions which arise from minimal (MIN) substitution into the OPE and CT parts of the *NN* interaction. Because, the nucleus is a composite object, there are additional "non-minimal" (NM) terms and this introduces LECs that are not contained in the nuclear potential. The five LECs governing the NM contributions are constrained with electromagnetic data; namely, the $A = 2 - 3$ magnetic moments, the deuteron magnetic form factor, and deuteron photodisintegration data [43].

References

1. R. Machleidt, D.R. Entem, Chiral effective field theory and nuclear forces. Phys. Rept. **503**, 1–75 (2011). https://doi.org/10.1016/j.physrep.2011.02.001
2. M. Piarulli, I. Tews, Local Nucleon-Nucleon and Three-Nucleon Interactions Within Chiral Effective Field Theory. Front. in Phys. **7**, 245 (2020). https://doi.org/10.3389/fphy.2019.00245
3. H.W. Hammer, S. König, U. van Kolck, Nuclear effective field theory: status and perspectives. Rev. Mod. Phys. **92**(2), 025004 (2020). https://doi.org/10.1103/RevModPhys.92.025004
4. M. Piarulli, L. Girlanda, R. Schiavilla, R. Navarro Pérez, J. E. Amaro, and E. Ruiz Arriola. Minimally nonlocal nucleon-nucleon potentials with chiral two-pion exchange including Δ resonances. Phys. Rev. **C91**(2), 024003 (2015). https://doi.org/10.1103/PhysRevC.91.024003
5. M. Piarulli, L. Girlanda, R. Schiavilla, A. Kievsky, A. Lovato, L.E. Marcucci, S.C. Pieper, M. Viviani, R.B. Wiringa, Local chiral potentials with Δ-intermediate states and the structure of light nuclei. Phys. Rev. C **94**(5), 054007 (2016). https://doi.org/10.1103/PhysRevC.94.054007

6. M. Piarulli et al., Light-nuclei spectra from chiral dynamics. Phys. Rev. Lett. **120**(5), 052503 (2018). https://doi.org/10.1103/PhysRevLett.120.052503
7. A. Baroni et al., Local chiral interactions, the tritium Gamow-Teller matrix element, and the three-nucleon contact term. Phys. Rev. C **98**(4), 044003 (2018). https://doi.org/10.1103/PhysRevC.98.044003
8. S. Pastore, R. Schiavilla, J.L. Goity, Electromagnetic two-body currents of one- and two-pion range. Phys. Rev. C **78**, 064002 (2008). https://doi.org/10.1103/PhysRevC.78.064002
9. S. Pastore, L. Girlanda, R. Schiavilla, M. Viviani, R.B. Wiringa, Electromagnetic Currents and Magnetic Moments in (chi)EFT. Phys. Rev. C **80**, 034004 (2009). https://doi.org/10.1103/PhysRevC.80.034004
10. S. Pastore, L. Girlanda, R. Schiavilla, M. Viviani, The two-nucleon electromagnetic charge operator in chiral effective field theory (χEFT) up to one loop. Phys. Rev. C **84**, 024001 (2011). https://doi.org/10.1103/PhysRevC.84.024001
11. M. Piarulli, L. Girlanda, L.E. Marcucci, S. Pastore, R. Schiavilla, M. Viviani, Electromagnetic structure of A = 2 and 3 nuclei in chiral effective field theory. Phys. Rev. C **87**(1), 014006 (2013). https://doi.org/10.1103/PhysRevC.87.014006
12. A. Baroni, L. Girlanda, S. Pastore, R. Schiavilla, M. Viviani, Nuclear Axial Currents in Chiral Effective Field Theory. Phys. Rev. C **93**(1), 015501 (2016). https://doi.org/10.1103/PhysRevC.93.049902, DOI: https://doi.org/10.1103/PhysRevC.93.015501, 10.1103/PhysRevC.95.059901. [Erratum: Phys. Rev. C 95, no.5,059901(2017)]
13. A. Baroni, L. Girlanda, A. Kievsky, L.E. Marcucci, R. Schiavilla, M. Viviani, Tritium β-decay in chiral effective field theory. Phys. Rev. C **94**(2), 024003 (2016). https://doi.org/10.1103/PhysRevC.94.024003, DOI: https://doi.org/10.1103/PhysRevC.95.059902. [Erratum: Phys. Rev. C 95, no.5,059902(2017)]
14. R. Schiavilla, A. Baroni, S. Pastore, M. Piarulli, L. Girlanda, A. Kievsky, A. Lovato, L. E. Marcucci, Steven C. Pieper, M. Viviani, and R. B. Wiringa. Local chiral interactions and magnetic structure of few-nucleon systems. Phys. Rev. C **99**, 034005 (2019). https://doi.org/10.1103/PhysRevC.99.034005
15. A. Bohr, B.R. Mottelson, Nuclear Structure (World Scientific Publishing Company, 1998). https://doi.org/10.1142/3530
16. G. Breit, E. Feenberg, The Possibility of the Same Form of Specific Interaction for All Nuclear Particles. Phys. Rev. **50**, 850–856 (1936). https://doi.org/10.1103/PhysRev.50.850
17. W. Heisenberg, On the structure of atomic nuclei. Z. Phys. **77**, 1–11 (1932). https://doi.org/10.1007/BF01342433
18. J.M.B. Kellogg, I.I. Rabi, N.F. Ramsey, J.R. Zacharias, An Electrical Quadrupole Moment of the Deuteron. Phys. Rev. **55**, 318–319 (1939). https://doi.org/10.1103/PhysRev.55.318
19. A.R. Edmonds, Angular momentum in quantum mechanics, vol. 4 (Princeton university press, 1996)
20. H. Yukawa, On the Interaction of Elementary Particles I. Proc. Phys. Math. Soc. Jap. **17**, 48–57 (1935). https://doi.org/10.1143/PTPS.1.1
21. R.B. Wiringa, V.G.J. Stoks, R. Schiavilla, An Accurate nucleon-nucleon potential with charge independence breaking. Phys. Rev. C **51**, 38–51 (1995). https://doi.org/10.1103/PhysRevC.51.38
22. J.L. Friar, B.F. Gibson, G.L. Payne, Recent progress in understanding trinucleon properties. Ann. Rev. Nucl. Part. Sci. **34**, 403–433 (1984). https://doi.org/10.1146/annurev.ns.34.120184.002155
23. S. Weinberg, Effective chiral Lagrangians for nucleon - pion interactions and nuclear forces. Nucl. Phys. B **363**, 3–18 (1991). https://doi.org/10.1016/0550-3213(91)90231-L
24. U. van Kolck, Few nucleon forces from chiral Lagrangians. Phys. Rev. C **49**, 2932–2941 (1994). https://doi.org/10.1103/PhysRevC.49.2932
25. C. Ordonez, L. Ray, U. van Kolck, The Two nucleon potential from chiral Lagrangians. Phys. Rev. C **53**, 2086–2105 (1996). https://doi.org/10.1103/PhysRevC.53.2086
26. E. Epelbaum, Walter Gloeckle, and Ulf-G. Meissner. Nuclear forces from chiral Lagrangians using the method of unitary transformation. 1. Formalism. Nucl. Phys. **A637**, 107–134 (1998). https://doi.org/10.1016/S0375-9474(98)00220-6

27. E. Epelbaum, A. Nogga, Walter Gloeckle, H. Kamada, Ulf G. Meissner, and H. Witala. Three nucleon forces from chiral effective field theory. Phys. Rev. **C66**, 064001 (2002). https://doi.org/10.1103/PhysRevC.66.064001
28. D.R. Entem, R. Machleidt, Accurate charge dependent nucleon nucleon potential at fourth order of chiral perturbation theory. Phys. Rev. C **68**, 041001 (2003). https://doi.org/10.1103/PhysRevC.68.041001
29. V. Bernard, E. Epelbaum, H. Krebs, U.G. Meissner, Subleading contributions to the chiral three-nucleon force II: Short-range terms and relativistic corrections. Phys. Rev. C **84**, 054001 (2011). https://doi.org/10.1103/PhysRevC.84.054001
30. J. Carlson, S. Gandolfi, F. Pederiva, Steven C. Pieper, R. Schiavilla, K. E. Schmidt, and R. B. Wiringa. Quantum Monte Carlo methods for nuclear physics. Rev. Mod. Phys. **87**, 1067 (2015). https://doi.org/10.1103/RevModPhys.87.1067
31. R. Navarro Pérez, J.E. Amaro, E. Ruiz Arriola, Coarse-grained potential analysis of neutron-proton and proton-proton scattering below the pion production threshold. Phys. Rev. **C88**(6), 064002 (2013). https://doi.org/10.1103/PhysRevC.88.064002, 10.1103/PhysRevC.91.029901. [Erratum: Phys. Rev.C91,no.2,029901(2015)]
32. R. Navarro Pérez, J. E. Amaro, and E. Ruiz Arriola. Coarse grained NN potential with Chiral Two Pion Exchange. Phys. Rev. C **89**(2), 024004 (2014). https://doi.org/10.1103/PhysRevC.89.024004
33. R. Navarro Perez, J.E. Amaro, E. Ruiz Arriola, Statistical error analysis for phenomenological nucleon-nucleon potentials. Phys. Rev. C **89**(6), 064006 (2014). https://doi.org/10.1103/PhysRevC.89.064006
34. J. Fujita, H. Miyazawa, Pion Theory of Three-Body Forces. Prog. Theor. Phys. **17**, 360–365 (1957). https://doi.org/10.1143/PTP.17.360
35. A. Gardestig, D.R. Phillips, How low-energy weak reactions can constrain three-nucleon forces and the neutron-neutron scattering length. Phys. Rev. Lett. **96**, 232301 (2006). https://doi.org/10.1103/PhysRevLett.96.232301
36. D. Gazit, S. Quaglioni, P. Navratil, Three-Nucleon Low-Energy Constants from the Consistency of Interactions and Currents in Chiral Effective Field Theory. Phys. Rev. Lett. **103**, 102502 (2009). https://doi.org/10.1103/PhysRevLett.103.102502
37. R. Schiavilla. . *unpublished*, 2017
38. T.S. Park, D.-P. Min, M. Rho, Chiral dynamics and heavy fermion formalism in nuclei. 1. Exchange axial currents. Phys. Rept. **233**, 341–395 (1993). https://doi.org/10.1016/0370-1573(93)90099-Y
39. T.-S. Park, D.-P. Min, M. Rho, Chiral Lagrangian approach to exchange vector currents in nuclei. Nucl. Phys. A **596**, 515–552 (1996). https://doi.org/10.1016/0375-9474(95)00406-8
40. N. Fettes, U.-G. Meissner, M. Mojzis, S. Steininger, The Chiral effective pion nucleon Lagrangian of order p**4. Annals Phys. **283**, 273–302 (2000). https://doi.org/10.1006/aphy.2000.6059. [Erratum: Annals Phys.288,249(2001)]
41. V. Bernard, N. Kaiser, U.-G. Meissner, Chiral dynamics in nucleons and nuclei. Int. J. Mod. Phys. E **4**, 193–346 (1995). https://doi.org/10.1142/S0218301395000092
42. G.B. King, A. Baroni, V. Cirigliano, S. Gandolfi, L. Hayen, E. Mereghetti, S. Pastore, M. Piarulli, Ab initio calculation of the β-decay spectrum of He6. Phys. Rev. C **107**(1), 015503 (2023). https://doi.org/10.1103/PhysRevC.107.015503
43. A. Gnech, R. Schiavilla, Magnetic structure of few-nucleon systems at high momentum transfers in a chiral effective field theory approach. Phys. Rev. C **106**(4), 044001 (2022). https://doi.org/10.1103/PhysRevC.106.044001

Chapter 3
Quantum Monte Carlo Methods

Abstract To study the electroweak properties of nuclei, one needs a method to compute nuclear wave functions and evaluate transition matrix elements. Thus, one must solve the Schrödinger equation for a many-body system coupled to the appropriate set of quantum numbers. Because of the complexity of the nuclear Hamiltonian, an analytic solution is intractable for most light systems and one must use numerical techniques to approach the problem. In this chapter, I will describe the two quantum Monte Carlo (QMC) approaches that I use to study electroweak processes; namely, Variational Monte Carlo (VMC) and Green's Function Monte Carlo (GFMC). In the QMC approach, one obtains wave functions and matrix elements through a stochastic procedure. This approach is a powerful tool in light systems, providing some of the most accurate calculations of properties for nuclei in the $A = 6$–12 mass range [1]. I will discuss the details, strengths, and limitations of the procedures in the following sections.

3.1 Variational Monte Carlo

To solve the many-body Schrödinger Equation $H\Psi(J^\pi; T, T_z) = E\Psi(J^\pi; T, T_z)$ for the appropriate angular momentum and parity (J^π), isospin (T), and isospin projection (T_z) quantum numbers with QMC is a two step process. The first step is the VMC method [2] and the heart of this approach is constructing a suitable ansatz for the many-body wave function. In strongly correlated fermion systems, an appropriate ansatz is,

$$\Psi = F\Phi\,, \tag{3.1}$$

where Φ is the unperturbed model wave function that encodes the quantum numbers and—in the case of fermions—the antisymmerty of the system, while F is a "correlation factor." The role of a correlation factor in the wave function for a system of particles interacting via a potential with strong, short-range repulsion is to reduce the relative wave function of two particles at short distances while imposing the appropriate long-range boundary conditions [3].

G. B. King, *Electroweak Structure and Reactions in Light Nuclei with Quantum Monte Carlo Methods*, Springer Theses, https://doi.org/10.1007/978-3-032-03191-4_3

The specific form of the VMC trial wave function used in calculations of finite nuclei has the form [4]

$$|\Psi_T\rangle = \mathcal{S}\prod_{i<j}^{A}\left[1 + U_{ij} + \sum_{k\neq i,j}^{A} \widetilde{U}_{ijk}^{TNI}\right]|\Psi_J\rangle\,, \tag{3.2}$$

with U_{ij} and $\widetilde{U}_{ijk}^{TNI}$ being two- and three-body correlation operators designed to reflect the impact of the nuclear interaction at short distances and $\mathcal{S}$ being the symmetrizer. The correlations arising for the *NN* and $3N$ forces induce behavior that goes beyond a single-particle picture. The component Ψ_J serves the role of the unperturbed model wave function and contains information about the symmetries and quantum numbers of the system. It is worth noting that this is a simplified version of the optimal variational wave function which factorizes the two- and three-body correlations as suggested by perturbation theory [2, 5–7]; however, it was found to be roughly twice as efficient computationally while still obtaining decent values for energies [4]. Thus, it is particularly effective as a starting point for GFMC calculations, which comprise the second step of the QMC approach and will be described in detail in Sect. 3.2.

The two-body correlation operator is designed to reflect the influence of the *NN* interaction at short distances has the form [1]

$$U_{ij} = \sum_{l=2}^{6} u^l(r_{ij})O_{ij}^l\,, \tag{3.3}$$

where O_{ij}^l are operators arising from the dominant one-pion exchange terms defined in Eq. 2.9 and $u^l(r_{ij})$ are pair correlations that come from solutions to a set of coupled differential equations written in the different ST channels of the *NN* system [5].

The three-body correlation operator is taken as

$$U_{ijk}^{TNI} = \sum_x \epsilon_x V_x(\widetilde{r}_{ij}, \widetilde{r}_{jk}, \widetilde{r}_{ik})\,, \tag{3.4}$$

where ϵ_x is a small, negative strength parameter and the interparticle spacing is scaled by an additional variational parameter y such that $\widetilde{r}_{ij} = yr_{ij}$ [1]. The correlation operator $\widetilde{U}_{ijk}^{TNI}$ is a truncated version of the three-body correlation operator in Eq. 3.4 that leaves out contributions from the central component of the $3N$ force [4].

Ψ_J is a Jastrow wave function that is fully anti-symmetric and has the quantum numbers of the state of interest. It is an improvement over a simple Slater determinant, as it includes information about the structure of clusters in the system. For finite nuclei, a typical Jastrow wave function Φ_J of a p-shell nucleus can be written as [1]

$$|\Phi_J\rangle = \mathcal{A}\prod_{i<j<k} f_{ijk}^c \prod_{i<j\leq 4} f_{ss}(r_{ij}) \prod_{k\leq 4<n\leq A} f_{sp}(r_{kn}) \sum_N \prod_{4<n<m\leq A} f_{pp}(r_{nm})|\Phi_N(1234;56\ldots A)\rangle. \tag{3.5}$$

Φ comes from an extreme shell model treatment with nucleons simply placed in the s- and p-shell and $\mathcal{A}$ is the anti-symmetrization operator. The particles 1 through 4 are taken to be in an alpha-particle cluster structure and the sum is over the $N = \binom{A}{4}$ partitions. $f_{ss}(r_{ij})$ is a function that encodes the correlations between two s-shell nucleons and, for p-shell nuclei, describes the alpha-cluster structure. The functions $f_{sp}(r_{ij})$ and $f_{pp}(r_{ij})$ encode correlations between s- and p-shell nucleons and amongst two nucleons in the p-shell, respectively. These cluster functions involve modifying $f_{ss}(r_{ij})$ with a long-range tail and a short-range Woods-Saxon factor. Finally, the f^c_{ijk} are the three-body correlations induced by the NN interaction. Detailed forms of these functions are given in Ref. [5].

The full Jastrow wave function Ψ_J is a linear combination of wave functions in the form given by Eq. 3.5 for the set of LS states coupled to the appropriate value of J^π. That is,

$$|\Psi_J\rangle = \sum_{LS[n]} \beta_{LS[n]} |\Phi_J^{LS[n]}\rangle\,, \tag{3.6}$$

where the $\beta_{LS[n]}$ are coefficients obtained via diagonalization and $[n]$ indicates the spatial symmetry of the wave function in the Young scheme [8, 9]. The Young pattern for k symmetric subgroups of nucleons is represented as $[n] = [n_1 n_2 \dots n_k]$ where the n_i's indicate the number of nucleons in the ith subgroup. Dropping the implied α cluster structure, $[n]$ for a p-shell nucleus may be found by grouping the nucleons by the three possible orbital angular momentum projections l_z. Assuming isospin symmetry, it is possible to have at most four nucleons in a given cluster. One can then find the allowed L values for a given Young diagram either by systematically raising and lowering the angular momentum projections starting from the state in the maximally allowed total orbital angular momentum projection $L_z = L$ or by using the symmetries of the corresponding Young Tableaux [10]. The allowed L values for different $[n]$ in the p-shell are given in Ref. [8]. As an example, consider the symmetry [2], which may only couple to $L = 2$ or 0 for symmetric identical particles. If one adds a third particle, it is possible either to form a [3] or [21] symmetry state. The former may only couple $L = 3$ or1 while the later can couple $L = 3, 2,$ or 1. The antisymmetric state, [11], may only couple $L = 1$ and the addition of the third particle may form a [21] or [111] symmetry. The ST values and the total J of the nucleus further restrict the allowed couplings so that only a subset are possible for a nucleus of given quantum numbers. For instance, using the notation $^{2S+1}L[n]$, the ^{6}Li$(1^+;0)$ ground state may be composed of $^3S[2]$, $^3D[2]$, and $^1P[11]$ states while the ^{6}He$(0^+;1)$ ground state is composed of a $^1S[2]$ and a $^3P[11]$ symmetry. The former is dominated by the $^3S[2]$ component which represents the strong $\alpha + d$ clustering of that state, while the dominant state of the latter is $^1S[2]$ [9].

The parameters that govern the correlation functions may be optimized by minimizing the energy expectation value,

$$E_V = \frac{\langle\Psi_V|H|\Psi_V\rangle}{\langle\Psi_V|\Psi_V\rangle} \geq E_0\,, \tag{3.7}$$

which, according to the variational principle, is an upper bound to the true ground state energy E_0. Metropolis Monte Carlo integration is used to evaluate the energy expectation value in the optimization procedure. As a result of the Monte Carlo sampling, there will be some amount of statistical uncertainty in the variational calculations of observable quantities.

3.2 Green's Function Monte Carlo

Diffusion Monte Carlo methods [11–14] are used to project the exact ground state of a given Hamiltonian via a propagation in imaginary time. Strictly speaking, DMC methods can only be applied for systems that depend on spatial degrees of freedom. Nuclear systems, as seen in Chap. 2, have spin and isospin dependencies that must be considered. The extension of DMC to include the relevant spin and isospin degrees of freedom is the GFMC method [15], which I use extensively for the study off-diagonal transition matrix elements in this work.

To understand how GFMC improves upon a trial wave function Ψ_T, one may start from the time-dependent Schrödinger Equation (with an energy offset E_T),

$$i\frac{\partial}{\partial t}|\Psi(t)\rangle = (H - E_T)|\Psi(t)\rangle\,, \tag{3.8}$$

which, for the initial condition $\Psi(0) = \Psi_T$, yields the solution

$$|\Psi(t)\rangle = e^{-i(H-E_T)t}|\Psi_T\rangle\,. \tag{3.9}$$

It is clear that one may make a change of variables from real time t to imaginary time $\tau = it$ to recast the Schrödinger Equation as a diffusion equation

$$-\frac{\partial}{\partial\tau}|\Psi(\tau))\rangle = (H - E_T)|\Psi(\tau)\rangle\,, \tag{3.10}$$

that has the solution

$$|\Psi(\tau)\rangle = e^{-(H-E_T)\tau}|\Psi_T\rangle\,. \tag{3.11}$$

Noting that one may expand any state in a complete orthonormal basis, the trial state can be written in terms of the eigenstates of the exact Hamiltonian ψ_i,

$$|\Psi_T\rangle = \sum_{i=1}^{\infty} c_i|\psi_i\rangle\,. \tag{3.12}$$

From Eq. 3.11, it is evident that if the energy offset E_T is taken to be the exact energy of the ground state E_0, then after an infinitely long propagation,

$$\lim_{\tau\to\infty} e^{-(H-E_0)\tau}|\Psi_T\rangle \propto c_0\psi_0. \tag{3.13}$$

In practice, this is achieved computationally by choosing E_T as close to E_0 as possible and acting on the state in several small imaginary time steps $\Delta\tau$,

$$|\Psi(\tau)\rangle = \left[e^{-(H-E_0)\Delta\tau}\right]^n |\Psi_T\rangle, \tag{3.14}$$

and looking for convergence in the energy or other observables of interest. In principle, so long as it is not orthogonal to the exact ground state of the Hamiltonian, any trial wave function used in this procedure will converge according to Eq. 3.13. Studies of various forms of trial wave function showed that the algorithm converges quickly when using the best VMC state Ψ_V as the input to GFMC.

When performing a GFMC calculation, one must consider the evolution of several sets of configurations $\mathbf{R}$, where $\mathbf{R}$ is used as short-hand to denote the set of $3A$ spatial coordinates and $2^A A!/(N!Z!)$ spin-isospin degrees of freedom. Thus, projecting Eq. 3.11 for a short imaginary time onto configuration space and using a complete set of orthonormal basis states $|\mathbf{R}'\rangle$, we obtain,

$$\Psi(\mathbf{R}, \tau + \Delta\tau) = \int d\mathbf{R}' \langle\mathbf{R}|e^{-(H-E_0)\Delta\tau}|\mathbf{R}'\rangle\Psi(\mathbf{R}', \tau) \tag{3.15}$$

$$= \int d\mathbf{R}' G(\mathbf{R}, \mathbf{R}', \Delta\tau)\Psi(\mathbf{R}', \tau)\,, \tag{3.16}$$

where I have used the definition the short-time propagator $G(\mathbf{R}, \mathbf{R}', \Delta\tau) = \langle\mathbf{R}|e^{-(H-E_0)\Delta\tau}|\mathbf{R}'\rangle$ and denoted the matrix element $\langle\mathbf{R}|\Psi(\tau)\rangle$ as $\Psi(\mathbf{R}, \tau)$. The propagation in imaginary time thus becomes a problem of sampling $G(\mathbf{R}, \mathbf{R}', \Delta\tau)$ with Monte Carlo techniques to evolve the configurations. A great amount of work has gone into obtaining the form of the propagator used in GFMC calculations and its description is beyond the scope of this thesis. I direct interested readers to Ref. [5] for more detail.

In principle, GFMC propagation should converge to the lowest energy state for a given set of quantum numbers. For calculations of physically wide states above the threshold for breakup into separate clusters, one expects the propagation of the energy to converge to the associated threshold. One typically observes in these instances a rapid decrease of the energy at short imaginary times before a slow decrease toward the threshold. The energy that is quoted for such a calculations is the value at small imaginary time ($\tau \sim 0.1$ MeV^{-1}) after the rapid improvement on the variational energy, but before the slow decrease to the threshold. Physically narrow excited states above, but very close, to the threshold can usually have the energy extracted in the same manner as the ground state calculation.

It is worth noting that one of the major challenges for GFMC (and generally for DMC simulations of many-fermion systems) is the so-called "sign problem." This challenge arises due to the fact that the Monte Carlo techniques used to obtain $\Psi(\tau)$ involve only local properties of the wave function, while antisymmetry is a global property. This gives wave functions with rapidly oscillating signs for large τ, resulting in large statistical errors for $\langle \mathcal{O} \rangle_M$ [4]. To deal with the fermion sign problem, constrained path methods [4, 16] exist to remove configurations which will contribute only noise to observables. If the exact ground state were known, one could in principle discard any walker which had zero overlap with the ground state; however, it is not known and can only be enforced approximately by discarding walkers having zero overlap with the VMC wave function according to a probabilistic scheme. This mitigates the sign problem at the cost of the energy now no longer approaches the true ground state from above [17]. To avoid bias from the constrained path propagation, a small number of unconstrained steps are carried out before the expectation values are evaluated [18].

3.2.1 Mixed Estimates

Ideally, for diagonal matrix elements of some operator $\mathcal{O}$, one would want to calculate the quantity

$$\langle O(\tau) \rangle = \frac{\langle \Psi(\tau) | \mathcal{O} | \Psi(\tau) \rangle}{\langle \Psi(\tau) | \Psi(\tau) \rangle} , \tag{3.17}$$

for all observables under study. An approximate expression is obtained by assuming the GFMC propagation is a small correction to the variational state. That is, we take $\Psi(\tau) = \Psi_V + \delta\Psi(\tau)$. Then, neglecting terms of order $[\delta\Psi(\tau)]^2$, the following approximation is obtained,

$$\begin{aligned} \langle O(\tau) \rangle &= \frac{\langle \Psi(\tau) | \mathcal{O} | \Psi(\tau) \rangle}{\langle \Psi(\tau) | \Psi(\tau) \rangle} \\ &\approx \langle O(\tau) \rangle_M + [\langle O(\tau) \rangle_M - \langle O \rangle_{\text{VMC}}] , \end{aligned} \tag{3.18}$$

where I have defined the "mixed estimate,"

$$\langle O(\tau) \rangle_M = \frac{\langle \Psi(\tau) | \mathcal{O} | \Psi_V \rangle}{\langle \Psi(\tau) | \Psi_V \rangle} . \tag{3.19}$$

If $\mathcal{O}$ commutes with H for the system, then Eq. 3.18 is exact. Otherwise, it is an approximation.

This was first generalized to off diagonal transitions in Ref. [20]. In these cases, the matrix element under consideration is of the form,

$$\langle \mathcal{O}(\tau)\rangle = \frac{\langle \Psi_f(\tau)|\mathcal{O}|\Psi_i(\tau)\rangle}{\sqrt{\langle \Psi_f(\tau)|\Psi_f(\tau)\rangle}\sqrt{\langle \Psi_i(\tau)|\Psi_i(\tau)\rangle}}. \tag{3.20}$$

This is then approximated as follows,

$$\begin{aligned}\langle \mathcal{O}(\tau)\rangle \approx \; & \frac{\langle \Psi_f(\tau)|\mathcal{O}|\Psi_i\rangle}{\sqrt{\langle \Psi_f(\tau)|\Psi_f\rangle}\sqrt{\langle \Psi_i|\Psi_i\rangle}} + \frac{\langle \Psi_f|\mathcal{O}|\Psi_i(\tau)\rangle}{\sqrt{\langle \Psi_f|\Psi_f\rangle}\sqrt{\langle \Psi_i|\Psi_i(\tau)\rangle}} \\ & - \frac{\langle \Psi_f|\mathcal{O}|\Psi_i\rangle}{\sqrt{\langle \Psi_f|\Psi_f\rangle}\sqrt{\langle \Psi_i|\Psi_i\rangle}},\end{aligned} \tag{3.21}$$

where I retain terms to leading order in $\delta\Psi$. Now, this may be recast in terms of quantities that one can evaluate in VMC and GFMC; namely, one may write,

$$\begin{aligned}\langle \mathcal{O}(\tau)\rangle \approx \; & \sqrt{\frac{\langle \Psi_f|\Psi_f(\tau)\rangle}{\langle \Psi_i|\Psi_i\rangle}}\frac{\langle \Psi_f(\tau)|\mathcal{O}|\Psi_i\rangle}{\langle \Psi_f(\tau)|\Psi_f\rangle} + \sqrt{\frac{\langle \Psi_i|\Psi_i(\tau)\rangle}{\langle \Psi_f|\Psi_f\rangle}}\frac{\langle \Psi_i(\tau)|\mathcal{O}^\dagger|\Psi_f\rangle}{\langle \Psi_i(\tau)|\Psi_i\rangle} \\ & - \frac{1}{2}\left[\sqrt{\frac{\langle \Psi_i|\Psi_i\rangle}{\langle \Psi_f|\Psi_f\rangle}}\frac{\langle \Psi_f|\mathcal{O}|\Psi_i\rangle}{\langle \Psi_i|\Psi_i\rangle} + \sqrt{\frac{\langle \Psi_f|\Psi_f\rangle}{\langle \Psi_i|\Psi_i\rangle}}\frac{\langle \Psi_i|\mathcal{O}^\dagger|\Psi_f\rangle}{\langle \Psi_f|\Psi_f\rangle}\right].\end{aligned} \tag{3.22}$$

The quantities $\langle \Psi_f(\tau)|\mathcal{O}|\Psi_i\rangle/\langle \Psi_f(\tau)|\Psi_f\rangle$ and $\langle \Psi_i(\tau)|\mathcal{O}^\dagger|\Psi_f\rangle/\langle \Psi_i(\tau)|\Psi_i\rangle$ are easily evaluated from the GFMC walk. The term in brackets is the VMC matrix element for the off diagonal transition averaged over the VMC walks for both the initial and final to improve the statistical uncertainty of the calculation. Finally, following Ref. [20], one further approximates the square root factors in the first line of Eq. 3.22 to obtain the standard expression used in GFMC calculations,

$$\begin{aligned}\langle \mathcal{O}(\tau)\rangle_{\mathrm{GFMC}} \approx \; & \sqrt{\frac{\langle \Psi_f|\Psi_f\rangle}{\langle \Psi_i|\Psi_i\rangle}}\frac{\langle \Psi_f(\tau)|\mathcal{O}|\Psi_i\rangle}{\langle \Psi_f(\tau)|\Psi_f\rangle} + \sqrt{\frac{\langle \Psi_i|\Psi_i\rangle}{\langle \Psi_f|\Psi_f\rangle}}\frac{\langle \Psi_i(\tau)|\mathcal{O}^\dagger|\Psi_f\rangle}{\langle \Psi_i(\tau)|\Psi_i\rangle} \\ & - \frac{1}{2}\left[\sqrt{\frac{\langle \Psi_i|\Psi_i\rangle}{\langle \Psi_f|\Psi_f\rangle}}\frac{\langle \Psi_f|\mathcal{O}|\Psi_i\rangle}{\langle \Psi_i|\Psi_i\rangle} + \sqrt{\frac{\langle \Psi_f|\Psi_f\rangle}{\langle \Psi_i|\Psi_i\rangle}}\frac{\langle \Psi_i|\mathcal{O}^\dagger|\Psi_f\rangle}{\langle \Psi_f|\Psi_f\rangle}\right].\end{aligned} \tag{3.23}$$

While the matrix elements propagated in GFMC typically display a clear convergence pattern for well-behaved wave functions, the mixed normalizations are less stable and have large statistical uncertainties. Thus, it is standard to adopt Eq. 3.23 to improve the error at the expense of a introducing a smaller contribution to the systematic uncertainty. Typically, one obtains the estimate at several steps in τ and computes the average after convergence is reached. An example of this propagation is seen in Fig. 3.1.

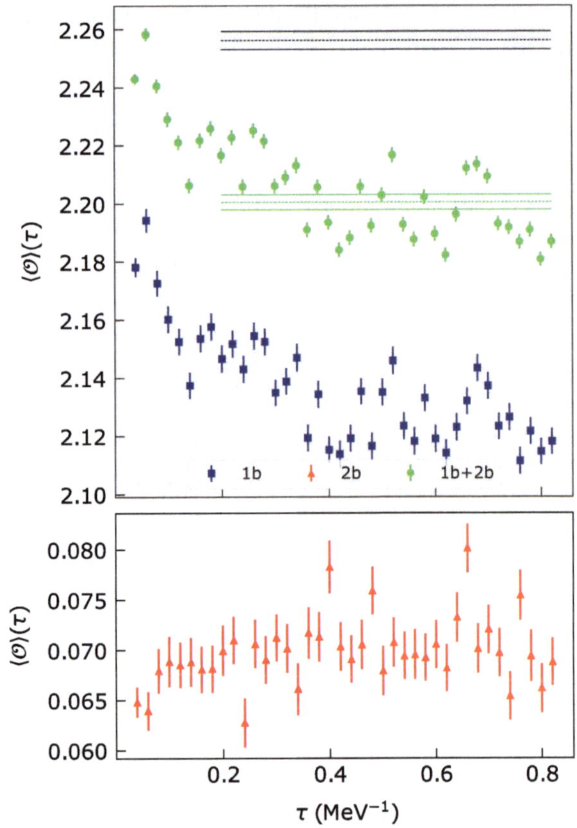

Fig. 3.1 Propagation of a ^{6}He→ ^{6}Li β-decay transition matrix element as function of imaginary time τ, based on the NV2+3-Ia Hamiltonian. Results with the one-body current at LO and currents beyond LO are indicated with 1b and 2b, respectively. Dashed and solid lines represent central values and associated error bars. Black dashed and solid lines denote the VMC results. Figure reproduced with permission from Ref. [19]

In theory, it would be possible to perform a calculation of $\langle O(\tau)\rangle$ where both the initial and final state wave functions are propagated in imaginary time. Such a calculation would require twice as long to compute as the mixed estimate. Further, each expectation value would also require its own GFMC propagation. Thus, if one intends to perform an order-by-order study of nuclear matrix elements, the problem quickly becomes very computationally intensive. Because of this, the standard approach for off-diagonal matrix elements in GFMC is the mixed estimate, which is a suitable approximation given the quality of the VMC wave functions used in nuclear calculations.

References

1. J. Carlson, S. Gandolfi, F. Pederiva, S.C. Pieper, R. Schiavilla, K.E. Schmidt, R.B. Wiringa, Quantum Monte Carlo methods for nuclear physics. Rev. Mod. Phys. **87**, 1067 (2015). https://doi.org/10.1103/RevModPhys.87.1067
2. R.B. Wiringa, Variational calculations of few-body nuclei. Phys. Rev. C **43**, 1585–1598 (1991). https://doi.org/10.1103/PhysRevC.43.1585
3. J.W. Clark, P. Westhaus, Method of Correlated Basis Functions. Phys. Rev. **141**, 833–857 (1966). https://doi.org/10.1103/PhysRev.141.833
4. R.B. Wiringa, S.C. Pieper, J. Carlson, V.R. Pandharipande, Quantum Monte Carlo calculations of A = 8 nuclei. Phys. Rev. C **62**, 014001 (2000). https://doi.org/10.1103/PhysRevC.62.014001
5. B.S. Pudliner, V.R. Pandharipande, J. Carlson, S.C. Pieper, R.B. Wiringa, Quantum Monte Carlo calculations of nuclei with A <= 7. Phys. Rev. C **56**, 1720–1750 (1997). https://doi.org/10.1103/PhysRevC.56.1720
6. K.M. Nollett, R.B. Wiringa, R. Schiavilla, A Six body calculation of the alpha deuteron radiative capture cross-section. Phys. Rev. C **63**, 024003 (2001). https://doi.org/10.1103/PhysRevC.63.024003
7. K.M. Nollett, Radiative alpha capture cross-sections from realistic nucleon-nucleon interactions and variational Monte Carlo wave functions. Phys. Rev. C **63**, 054002 (2001). https://doi.org/10.1103/PhysRevC.63.054002
8. A. Bohr, B.R. Mottelson, *Nuclear Structure* (World Scientific Publishing Company, 1998). https://doi.org/10.1142/3530
9. R.B. Wiringa, Pair counting, pion-exchange forces, and the structure of light nuclei. Phys. Rev. C **73**, 034317 (2006). https://doi.org/10.1103/PhysRevC.73.034317
10. A. Zee, *Group theory in a nutshell for physicists*, vol. 17. (Princeton University Press, 2016)
11. M.H. Kalos, Monte carlo calculations of the ground state of three- and four-body nuclei. Phys. Rev. **128**, 1791–1795 (Nov1962)
12. M.H. Kalos, Energy of a boson fluid with lennard-jones potentials. Phys. Rev. A **2**, 250–255 (Jul1970)
13. M.H. Kalos, D. Levesque, L. Verlet, Helium at zero temperature with hard-sphere and other forces. Phys. Rev. A **9**, 2178–2195 (May1974)
14. D.M. Ceperley, B.J. Alder, Ground state of the electron gas by a stochastic method. Phys. Rev. Lett. **45**, 566–569 (Aug1980)
15. J. Carlson, Green's function Monte Carlo study of light nuclei. Phys. Rev. C **36**, 2026–2033 (1987). https://doi.org/10.1103/PhysRevC.36.2026
16. S. Zhang, J. Carlson, J.E. Gubernatis, Constrained path monte carlo method for fermion ground states. Phys. Rev. B **55**, 7464–7477 (1997). https://doi.org/10.1103/PhysRevB.55.7464
17. J. Carlson, J.E. Gubernatis, G. Ortiz, S. Zhang, Issues and observations on applications of the constrained-path monte carlo method to many-fermion systems. Phys. Rev. B **59**, 12788–12798 (1999). https://doi.org/10.1103/PhysRevB.59.12788
18. S. Gandolfi, D. Lonardoni, A. Lovato, M. Piarulli, Atomic nuclei from quantum Monte Carlo calculations with chiral EFT interactions (2020)
19. G.B. King, L. Andreoli, S. Pastore, M. Piarulli, R. Schiavilla, R.B. Wiringa, J. Carlson, S. Gandolfi, Chiral effective field theory calculations of weak transitions in light nuclei. Phys. Rev. C **102**, 025501 (Aug2020)
20. M. Pervin, S.C. Pieper, R.B. Wiringa, Quantum Monte Carlo calculations of electroweak transition matrix elements in A = 6,7 nuclei. Phys. Rev. C **76**, 064319 (2007). https://doi.org/10.1103/PhysRevC.76.064319

Chapter 4
Gamow-Teller β-Decay in Light Nuclei

Abstract Though they are limited to relatively light nuclei due to the complexity of the computations, microscopic approaches—like QMC—are highly accurate where they are applicable. This makes light nuclei valuable for testing models of nuclear structure and electroweak currents. Data on electroweak transitions abound for these systems and this fact—combined with strict selection rules in the limit of vanishing momentum transfer—allows for stringent tests of the vector and axial currents that will be used to analyze new physics searches. In this section, I will discuss several calculations of electroweak observables at low-momentum transfer ($q \rightarrow 0$) using QMC techniques, focusing on purely axial processes. The calculations discussed include both model validation and the prediction of an experimentally relevant SM quantity for BSM searches.

4.1 Gamow-Teller β-Decay in Light Nuclei

Nuclear β-decay is a process that converts a neutron inside the nucleus to a proton with the emission of an electron and an electron antineutrino in the final state. It is also possible for nuclei to undergo β^+-decay, where a proton is converted to a neutron with the emission of a positron and an electron neutrino. From this point on, I will refer to both processes as β-decay. Electron capture—where a nucleus absorbs an electron to convert a proton into a neutron—is theoretically similar due to the small momentum transfer in this process. In the limit $q \rightarrow 0$, the electron capture rate can be extracted by computing the same matrix elements as one would for β-decay.

"Gamow-Teller (GT)" β-decay transitions—where the outgoing leptons couple to total spin 1–are of particular interest and have been instrumental as a constraint of axial current operators. Conventionally, the GT reduced matrix element $M_{\rm GT}$ is defined in terms of the longitudinal multipole $L_1(q)$ (see Eq. B.74),

G. B. King, *Electroweak Structure and Reactions in Light Nuclei with Quantum Monte Carlo Methods*, Springer Theses, https://doi.org/10.1007/978-3-032-03191-4_4

$$M_{\rm GT} = -i(-)^{J_i-M}\frac{\sqrt{2J_f+1}}{g_A}L_1(0) \tag{4.1}$$

$$= \frac{1}{g_A}\frac{\sqrt{2J_f+1}}{c^M_{J_fJ_iL}}\langle J_f M|\hat{\mathbf{z}}\cdot\mathbf{j}^\dagger_\pm(\mathbf{q}\to 0)|J_i M\rangle\,, \tag{4.2}$$

where g_A is the nucleon axial coupling constant and the current that raises or lowers isospin by one unit is defined as $\mathbf{j}^\dagger_\pm = \mathbf{j}^\dagger_x \pm i\mathbf{j}^\dagger_y$, with the Cartesian indices indicating isospin components. To provide a more concrete intuition of what this object is, it is useful to see its connection to experiment. The "ft value"—that is the product of the statistical rate function f, which depends on measurements of the end-point energy of the decay, and the partial half-life t of the transition that comes from measurements of the branching ratio and the half-life—is directly related to the reduced matrix element for β-decay [1, 2],

$$M_\beta = \sqrt{2J_i+1}\sqrt{\frac{K}{ft}}\,, \tag{4.3}$$

where K is a constant kinematic factor and,

$$M_\beta = M_{\rm F} + g_A M_{\rm GT}\,, \tag{4.4}$$

with the "Fermi" matrix element $M_{\rm F} \propto C_0(0)$ (see Eq. B.75). In a purely GT transition, $M_{\rm F}=0$, and the rate is directly related to $M_{\rm GT}$.

4.1.1 GT Transitions with QMC Techniques

A long-standing puzzle in the theory of GT transitions was the so-called "g_A quenching problem". Previous evaluations of $M_{\rm GT}$ in the shell model using one-body current operators required a quenching factor in order to reduce the theoretical predictions of $M_{\rm GT}$ to the values extracted from experiment [2]. Because the shell model is a mean field picture, it was unclear if this discrepancy originated from missing correlations in the wave function or from the omission of higher order many-body axial currents. While great effort went into understanding and establishing the importance of correlations in quenching spin-isospin responses in nuclei [3–7], the role of two-body currents in these processes have only relatively recently come under study by the theory community.

Early efforts to obtain $M_{\rm GT}$ starting from interacting nucleon degrees of freedom were performed with phenomenological models [8, 9]. More recently, QMC calculations were performed giving the first microscopic calculation of this quantity based on currents derived from χEFT and wave functions computed with the phenomenological AV18+IL7 potential, a framework referred to as the "hybrid approach" [10]. The authors of this study found, consistently with the intuition developed in works such as those of Refs. [3–7], that correlations in the nuclear wave function drive quenching of M_{GT}. In fact, for QMC calculations of transitions involving $6 \le A \le 10$ nuclei, they were the primary driver of quenching and the effect of many-body currents was always a small (~2% to 3%), additive contribution to M_{GT}.

Following the pioneering calculations of Ref. [10], a similar calculation was performed in light nuclei using the no-core shell model (NCSM) wave functions and in heavier nuclei using coupled cluster (CC) techniques [11]. This calculation used a potential and a set of currents derived in the same χEFT framework. The calculations confirmed that, in light systems, the quenching is driven by correlations; however, there was a discrepancy in the sign of the two-body correction, which sometimes provided an additional quenching on top of the correlations. The two-body current quenching was even more pronounced in CC calculations in heavier systems using a similarity renormalization group [12] evolved potential and current operators. The authors of Ref. [11] attributed the consistent enhancement of the two-body contribution in the QMC calculation to the hybrid nature of the first approach and speculated a mismatch in the short-range dynamics arising when one fits the unknown LEC c_D using AV18+IL7 wave functions. Thus, the discrepancy between theoretical and experimental values of M_{GT} had yet to be fully understood.

To better understand the disagreement in the nuclear theory approaches, I performed novel QMC calculations with the NV2+3 potential and consistent electroweak currents that were first presented in Ref. [13]. For $6 \le A \le 10$ nuclei, VMC calculations were performed using all eight of the NV2+3 model classes. These results are presented in Table 4.1. The results are broken down according to their order in the tree-level expansion of the current; namely, I provide the results for one-body axial current at LO, the one-body relativistic corrections (N2LO-RC) plus the two-body currents of one-pion range (N2LO-Δ) at N2LO, and the contributions at N3LO. The N3LO contribution is split into its two constituent parts, N3LO-OPE and N3LO-CT, corresponding to the one-pion range and contact currents, respectively. The cumulative contributions are given in the column labelled "Total," while the sum of the subleading contributions is labeled "Total–LO". Experimental data from Refs. [2, 14–17] are reported for comparison.

There is a qualitative agreement between VMC results based on different nuclear Hamiltonians. In particular, for the $A = 3, 6, 7$ and 10 systems, the LO contribution is 97% of the total matrix elements while subleading orders correct M_{GT} by $\lesssim 3\%$ correction. This correction is always constructive to the LO contribution for all cases being considered, except for the $A = 10$ transition. This outlying case exhibits an interesting model dependence; particularly, all of the starred models are quenched by the inclusion of subleading currents, while the unstarred models (expect for the NV2+3-IIb) receive an additive contribution.

The $A = 8$ transitions were also unique compared to the other cases studied. In these transitions, the LO current was strongly quenched relative to the experimental result. Initially, in Ref. [13], this behavior was attributed to the fact that the initial and final VMC wave functions are characterized by different dominant spatial symmetries. Therefore, the overlap between the initial and final wave functions is smaller

Table 4.1 M_{GT} in $A = 6, 7, 8$, and 10 nuclei obtained with chiral axial currents and VMC wave functions for the the NV2+3-Ia/b and NV2+3-IIa/b (NV2+3-Ia/b* and NV2+3-IIa/b*) model classes. Columns labeled with LO, N2LO-(RC+Δ), N3LO-OPE, and N3LO-CT refer to the contributions from Table 2.1 with explicit expressions as summarized in Sect. A.4. The sum of all contributions is reported in the column labeled "Total", while the subleading correction only is given under "Total–LO". Experimental values from Refs. [2, 14–17] are listed under "Expt." The transitions and the dominant spatial symmetries of the relevant VMC wave functions are given in the first column. Statistical errors associated with the Monte Carlo integration are not shown, but are below 1%

Transition	Model	LO	N2LO-(RC+Δ)	N3LO-OPE	N3LO-CT	Total–LO	Total	Expt.
^{6}He$(0^+;1) \to {}^{6}$Li$(1^+;0)$	Ia (Ib)	2.200 (2.254)	0.022 (0.056)	0.039 (0.064)	–0.005 (–0.068)	0.056 (0.052)	2.256 (2.306)	2.1609(40)
[42]→[42]	IIa (IIb)	2.207 (2.212)	0.027 (0.043)	0.043 (0.055)	–0.034 (–0.082)	0.036 (0.016)	2.243 (2.228)	
	Ia* (Ib*)	2.192 (2.256)	0.021 (0.056)	0.038 (0.063)	–0.054 (–0.097)	0.005 (0.022)	2.197 (2.279)	
	IIa* (IIb*)	2.202 (2.218)	0.027 (0.044)	0.043 (0.056)	–0.057 (–0.090)	0.014 (0.010)	2.216 (2.228)	
^{7}Be$(\frac{3}{2}^-;\frac{1}{2}) \to {}^{7}Li(\frac{3}{2}^-;\frac{1}{2})$	Ia (Ib)	2.317 (2.294)	0.099 (0.162)	0.076 (0.118)	–0.010 (–0.148)	0.165 (0.133)	2.482 (2.427)	2.3556(47)
[43]→[43]	IIa (IIb)	2.293 (2.309)	0.102 (0.153)	0.078 (0.113)	–0.070 (–0.190)	0.110 (0.076)	2.403 (2.385)	
	Ia* (Ib*)	2.327 (2.307)	0.098 (0.161)	0.076 (0.117)	–0.121 (–0.212)	0.053 (0.066)	2.380 (2.373)	
	IIa* (IIb*)	2.296 (2.316)	0.103 (0.154)	0.078 (0.114)	–0.120 (–0.210)	0.061 (0.058)	2.357 (2.374)	
^{7}Be$(\frac{3}{2}^-;\frac{1}{2}) \to {}^{7}Li(\frac{1}{2}^-;\frac{1}{2})$	Ia (Ib)	2.157 (2.119)	0.066 (0.122)	0.063 (0.100)	–0.009 (–0.125)	0.121 (0.096)	2.278 (2.215)	2.1116(57)
[43]→[43]	IIa (IIb)	2.128 (2.145)	0.069 (0.111)	0.065 (0.095)	–0.059 (–0.162)	0.074 (0.044)	2.202 (2.189)	
	Ia* (Ib*)	2.158 (2.124)	0.065 (0.119)	0.063 (0.099)	–0.103 (–0.180)	0.025 (0.038)	2.183 (2.162)	
	IIa* (IIb*)	2.131 (2.148)	0.067 (0.111)	0.064 (0.095)	–0.101 (–0.178)	0.030 (0.028)	2.161 (2.176)	

(continued)

Table 4.1 (continued)

Transition	Model	LO	N2LO-(RC+Δ)	N3LO-OPE	N3LO-CT	Total–LO	Total	Expt.
$^{8}\mathrm{Li}(2^{+};1)\rightarrow{}^{8}\mathrm{Be}(2^{+};0)$	Ia (Ib)	0.147 (0.092)	0.032 (0.028)	0.011 (0.011)	–0.001 (–0.014)	0.041 (0.031)	0.188 (0.123)	0.286 Ref. [16]
[431]→[44]	IIa (IIb)	0.144 (0.101)	0.031 (0.033)	0.010 (0.011)	–0.008 (–0.019)	0.033 (0.025)	0.177 (0.126)	0.191 Ref. [17]
	Ia* (Ib*)	0.148 (0.099)	0.032 (0.033)	0.010 (0.012)	–0.016 (–0.020)	0.026 (0.025)	0.174 (0.124)	
	IIa* (IIb*)	0.124 (0.121)	0.032 (0.037)	0.010 (0.013)	–0.014 (–0.023)	0.028 (0.027)	0.152 (0.148)	
$^{8}\mathrm{B}(2^{+};1)\rightarrow{}^{8}\mathrm{Be}(2^{+};0)$	Ia (Ib)	0.146 (0.092)	0.032 (0.032)	0.011 (0.011)	–0.001 (–0.014)	0.042 (0.030)	0.188 (0.122)	0.269(20)
[431]→[44]	IIa (IIb)	0.144 (0.102)	0.031 (0.033)	0.010 (0.011)	–0.008 (–0.019)	0.033 (0.026)	0.177 (0.128)	
	Ia* (Ib*)	0.148 (0.098)	0.032 (0.034)	0.010 (0.012)	–0.016 (–0.020)	0.026 (0.025)	0.174 (0.123)	
	IIa* (IIb*)	0.126 (0.118)	0.032 (0.037)	0.010 (0.013)	–0.014 (–0.022)	0.028 (0.027)	0.154 (0.145)	
$^{8}\mathrm{He}(0^{+};1)\rightarrow{}^{8}\mathrm{Li}(1^{+};1)$	Ia (Ib)	0.386 (0.363)	0.030 (0.034)	0.009 (0.012)	–0.001 (–0.014)	0.038 (0.032)	0.424 (0.396)	0.512(6)
[422]→[431]	IIa (IIb)	0.465 (0.370)	0.032 (0.034)	0.012 (0.011)	–0.009 (–0.017)	0.035 (0.028)	0.500 (0.398)	
	Ia* (Ib*)	0.362 (0.377)	0.031 (0.034)	0.009 (0.014)	–0.010 (–0.022)	0.029 (0.026)	0.391 (0.402)	
	IIa* (IIb*)	0.481 (0.364)	0.033 (0.035)	0.012 (0.012)	–0.017 (–0.019)	0.029 (0.028)	0.510 (0.391)	
$^{10}\mathrm{C}(0^{+};1)\rightarrow{}^{10}\mathrm{B}(1^{+};0)$	Ia (Ib)	1.940 (2.118)	0.003 (0.026)	0.044 (0.068)	–0.006 (–0.081)	0.041 (0.011)	1.981 (2.129)	1.8331(34)
[442]→[442]	IIa (IIb)	2.157 (2.176)	-0.002 (0.026)	0.046 (0.072)	–0.042 (–0.120)	0.002 (–0.022)	2.159 (2.154)	
	Ia* (Ib*)	2.015 (2.123)	0.024 (0.022)	0.059 (0.069)	–0.119 (–0.117)	–0.037 (–0.028)	1.978 (2.095)	
	IIa* (IIb*)	2.071 (2.143)	0.015 (0.031)	0.061 (0.072)	–0.126 (–0.123)	–0.050 (–0.020)	2.021 (2.123)	

relative to cases in which both states display the same dominant spatial symmetry. As a consequence, in these cases, the LO term is only about $\approx$ 40–50% of the total matrix element, with two-body currents providing a large additive correction. The inclusion of subleading currents improves the agreement with experiment, but the correction is insufficient to describe the data. Because of the reduced overlap between dominant components in the wave functions, these matrix elements are particularly sensitive to small components, which are poorly constrained and model dependent. Improved VMC wave functions could lead to better agreement with the data; however, recent work has made it clear that these transitions are experimentally complicated, as well. Ref. [18] comments on a possible intruder 2^+ state in ^{8}Be at $\sim$9 MeV that would also be accessible via β-decays from the ^{8}Li and ^{8}B ground states. These observations highlight the necessity for further investigation of these transitions with accurate many-body frameworks and future GFMC studies to address the possible presence and nature of the intruder state. Experimental confirmation or disproof of this state would also be highly beneficial to the nuclear theory community.

Common to all of the transitions was that the CT current at N3LO always quenches the LO term, while the currents of one-pion range at N2LO and N3LO enhance $M_{\rm GT}$. The main difference between the unstarred and starred models having the same NV2 potential is the size of the contact contribution. That the two models provided different behavior is unsurprising, as the LEC z_0 is proportional to the unknown $3N$ force LEC c_D (as shown in Eq. A.107 and discussed in Chap. 2) while the long-range contributions depend on known LECs. Since z_0 governs the strength of this term and the two fits provide different values of c_D (see Table A.5), one should expect the contribution to differ. Thus, how one approaches the fit to the three-body force has an impact on axial transitions. It is worth stressing, however, that this is still a modest correction to the overall $M_{\rm GT}$ in most cases, and although it generates a significant model uncertainty on this term, it makes a relatively small contribution.

To further investigate the difference of the three-body force dependence, I performed more accurate GFMC calculations with two model classes sharing the same NV2 potential but differing in the $3N$ LECs; specifially, the NV2+3-Ia/Ia* models. I present the resultant LO, total, and total subleading ("Total —LO") matrix elements compared to experiment in Table 4.2. Comparison to the hybrid results of Ref. [10] are included where applicable. This information is visualized in Fig. 4.1.

GFMC propagation reduces the VMC results by $\lesssim$ 4% in all selected transitions (save for the $A = 10$ transition obtained with the NV2+3-Ia* model). The agreement with the data, after the inclusion of two-body currents, is at the $\approx$ 2% ($\lesssim$ 2%) level for the $A = 6$ transition with the NV2+3-Ia (NV2+3-Ia*) model; and at the $\lesssim$ 4% ($\lesssim$ 1%) level for the $A = 7$ cases with the NV2+3-Ia (NV2+3-Ia*) model. These results are also in good agreement with those obtained for the same transitions in Ref. [10]. The NV2+3 models provide a better agreement with the data for the $A = 6$ primarily because the LO term is 2% smaller than obtained using AV18+IL7 model.

The largest dependence on the nuclear model occurs for the $A = 10$ transition. This is true both at the VMC level (see Table 4.1) and at the GFMC level in Table 4.2. As was iterated earlier, the VMC values display a large cutoff dependence (models a vs models b), as well as a large sensitivity to the energy range of scattering used

Table 4.2 M_{GT} in $A = 6$, 7, 8, and 10 nuclei obtained with chiral axial currents and GFMC (VMC) wave functions corresponding to the NV2+3-Ia and NV2+3-Ia* model classes. I show the results corresponding to the one-body current at LO ("LO") and the sum of all subleading corrections ("Total–LO"). The cumulative contribution under "Total" is to be compared with the experimental data [2, 14–17] reported in the last row. Where available, I show the results from the study of Ref. [10] based on the AV18+IL7 nuclear Hamiltonian for comparison. Statistical errors associated with the Monte Carlo integration are not shown, but are below 1%. Transitions for the $A = 8$ systems are affected by an additional systematic error of $\approx 5\%$, as explained in the text

Transition	Model	s.s.	LO	Total–LO	Total	Expt.
^{6}He(0$^+$;1)→ ^{6}Li(1$^+$;0)	Ia	[42]→[42]	2.130(2.200)	0.070(0.056)	2.201(2.256)	2.1609(40)
	Ia*		2.107(2.192)	0.011(0.005)	2.118(2.197)	
	Ref. [10]		2.168(2.174)	0.037(0.030)	2.205(2.211)	
^{7}Be($\frac{3}{2}^-$;$\frac{1}{2}$)→^{7}Li($\frac{3}{2}^-$;$\frac{1}{2}$)	Ia	[43]→[43]	2.273(2.317)	0.164(0.165)	2.440(2.482)	2.3556(47)
	Ia*		2.286(2.327)	0.052(0.053)	2.338(2.380)	
	Ref. [10]		2.294(2.334)	0.061(0.050)	2.355(2.384)	
^{7}Be($\frac{3}{2}^-$;$\frac{1}{2}$)→^{7}Li($\frac{1}{2}^-$;$\frac{1}{2}$)	Ia	[43]→[43]	2.065(2.157)	0.103(0.121)	2.168(2.278)	2.1116(57)
	Ia*		2.061(2.158)	0.009(0.025)	2.070(2.183)	
	Ref. [10]		2.083(2.150)	0.046(0.046)	2.129(2.196)	
^{8}Li(2$^+$;1)→^{8}Be(2$^+$;0)	Ia	[431]→[44]	0.074(0.147)	0.029(0.041)	0.103(0.188)	0.284 Ref. [16]
	Ia*		0.096(0.148)	0.025(0.026)	0.120(0.174)	0.190 Ref. [17]
^{8}B(2$^+$;1)→^{8}Be(2$^+$;0)	Ia	[431]→[44]	0.091(0.146)	0.035(0.042)	0.125(0.188)	0.269(20)
	Ia*		0.102(0.148)	0.024(0.026)	0.126(0.174)	
^{8}He(0$^+$;1)→^{8}Li(1$^+$;1)	Ia	[422]→[431]	0.262(0.386)	0.040(0.038)	0.302(0.424)	0.512(6)
	Ia*		0.297(0.362)	0.025(0.029)	0.322(0.391)	
^{10}C(0$^+$;1)→^{10}B(1$^+$;0)	Ia	[442]→[442]	1.928(1.940)	0.050(0.041)	1.978(1.981)	1.8331(34)
	Ia*		2.086(2.015)	–0.031(–0.037)	2.055(1.978)	
	Ref. [10]		2.032(2.062)	0.016(0.015)	2.048(2.077)	

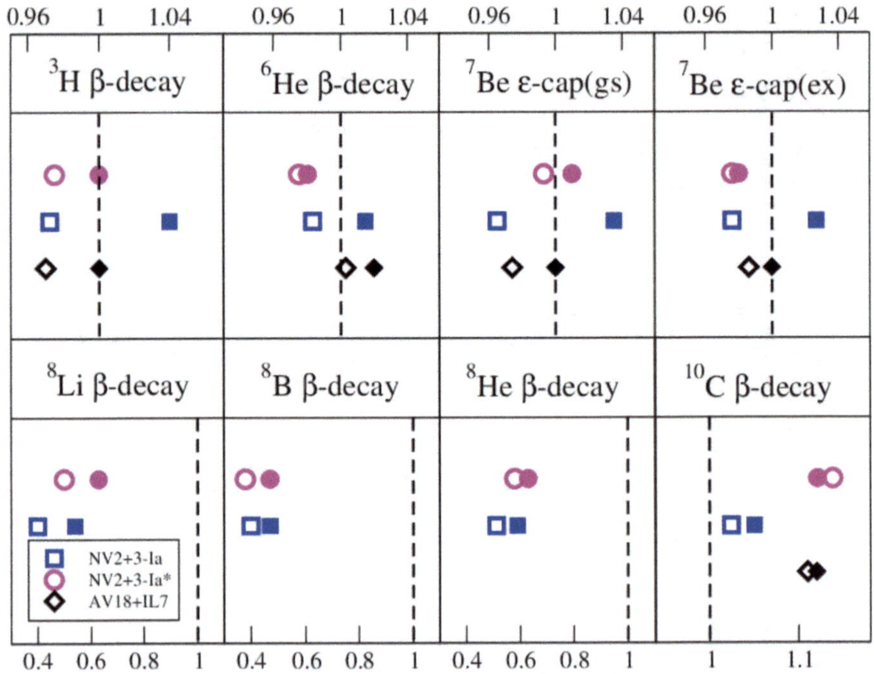

Fig. 4.1 Ratios of GFMC to experimental values of the M_{GT} in the ^{3}H, ^{6}He, ^{7}Be, ^{8}B, ^{8}Be, ^{8}He and ^{10}C weak transitions. Theory predictions corresponding to the χEFT axial current at LO (empty symbols) and up to N3LO (filled symbols) obtained with the NV2+3-Ia and NV2+3-Ia* models are shown with squares and pink circls, respectively. Results from Ref. [10] with the AV18+IL7 nuclear Hamiltonian and N4LO currents from Ref. [19] are represented with black diamonds. Results for the ^{3}H weak transition were reported in Ref. [20]. Figure reproduced with permission from Ref. [13]

to fit the NV2 (either I or II). The AV18+IL7 results of Ref. [10] lie between models Ia and Ia*. The structure of ^{10}B makes this a challenging nucleus to model, which contributes to the large differences in the two NV2+3 results. In this nucleus, there are two nearby ($\Delta E \sim 1$ MeV) $J^\pi = 1^+$ excited states, with the lower one having a predominantly $^3\mathrm{S}_1[442]$ spatial symmetry (using the convention for denoting the Young pattern in Sect. 3.1) and the higher-energy one being dominated by $^3\mathrm{D}_1[442]$. The transition from the $^{10}\mathrm{C}(0^+)$ state, which is predominantly $^1\mathrm{S}_0[442]$, is large in the $S \to S$ components, but about five times smaller in the $S \to D$ components. This makes the GT matrix element particularly sensitive to the exact mixing of the $^3\mathrm{S}_1$ and $^3\mathrm{D}_1$ components in the two $^{10}\mathrm{B}(1^+)$ states produced by a given Hamiltonian. Thus, it was concluded that none of the interaction models in Table 4.2 gets the mixing of the states exactly correct. In particular, results based on the NV2+3-Ia and NV2+3-Ia* interactions over-predict the data by $\sim$7% and 12%, respectively, which roughly indicates the range of theoretical uncertainty on the estimates.

This work represented the first QMC prediction of M_{GT} for $A = 8$ transitions to include the contribution from two-body axial currents. The extrapolation of the

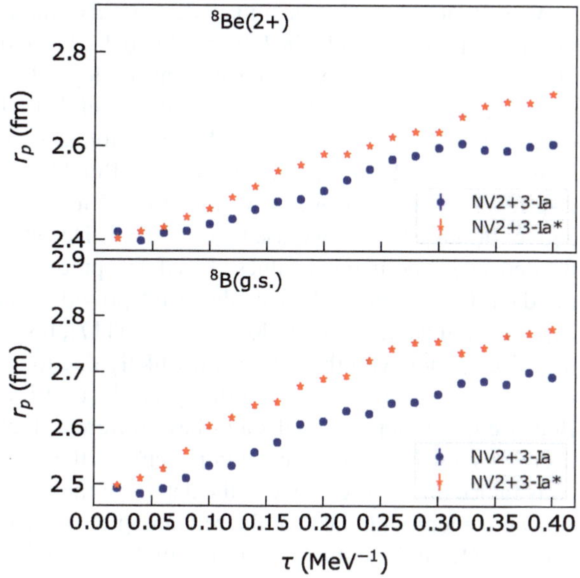

Fig. 4.2 GFMC propagation of the point-proton radius of the first excited state of ^{8}Be (upper panel) and ^{8}B (lower panel) based on the NV2+3-Ia and NV2+3-Ia* Hamiltonians. Figure reproduced with permission from Ref. [13]

GFMC matrix elements for transitions involving the $(J^\pi, T) = (2^+, 0)$ state of ^{8}Be and the ground state of ^{8}B had to be treated differently than other cases. For these two states, it was observed that the binding energy, the magnitude of the quadrupole moment, and the point-proton radius all increased monotonically with imaginary time. An example is shown in Fig. 4.2, which displays the point-proton radii of the two nuclear states. In Ref. [13], this behavior was interpreted as an indication that the resonant excited state of ^{8}Be is separating into two α's, while ^{8}B is breaking into $p+^7$Be. In the case of ^{8}Be, this issue has been addressed already in Refs. [21–23]. Here, I employed similar techniques to treat these systems and extract matrix elements from the GFMC data. In particular, I note that the ground-state energy of ^{8}Be drops very quickly as the imaginary time increases and reaches stability around $\tau \approx 0.1$ MeV^{-1}. The transition matrix elements involving the two dissolving states are extracted assuming that $\tau \approx 0.1$ MeV^{-1} is also the point at which spurious contamination in the nuclear wave functions is removed by the GFMC propagation. I performed the average in a small interval around this point, typically between 0.06 and 0.14 MeV^{-1}. Calculations involving these two systems are clearly affected by a systematic error which is roughly estimated by studying sensitivity of the extracted matrix elements with respect to variations in the imaginary time interval used in the averaging. Such a procedure generates an additional uncertainty of $\approx 5\%$, as noted in Table 4.2.

As discussed for the VMC, the LO value of the $A = 8$ transitions is suppressed, leading to only ≈50% and ≈40% of the experimental values for the ^{8}B → ^{8}Be and ^{8}He → ^{8}Li transitions. Two-body currents provide an enhancement of about 20% to 30% , but this remains insufficient to reach agreement with the experimental data. As previously noted, these transitions are not only challenging for theory, but also for experiment. For example, in Tables 4.1 and 4.2 I also quote two results for the RME of the ^{8}Li$\rightarrow$ ^{8}Be decay obtained from the log(ft)-values of Refs. [16] and [17] using Eq. 4.3 and the value $K = 6139$ from Ref. [2]. The Fermi transition strength is small enough in this case to treat the transition as purely axial. The associated $M_{\rm GT}$ values are $M_{\rm GT}$(Ref. [16])=0.284 and $M_{\rm GT}$ (Ref. [17])=0.190. We note that Refs. [15, 24], report a different overall factor of 6147 instead of 6139 in the formula given above. In our estimate we used $g_A = 1.2723$. Despite this additional uncertainty in the deduced experimental values, the GFMC predictions still severely underestimate the data. For example, the calculated Ia* provides only $\approx 40\%$ and $\approx 60\%$ of the experimental values given in Refs. [16] and [17], respectively. For the decay of ^{8}He, this discrepancy with the data is most likely due to the presence of three low-lying $(J^\pi, T) = (1^+, 1)$ states in the decay product, ^{8}Li, at $\approx$1, $\approx$3.2, and $\approx$5.4 MeV. Because the initial and final states have different dominant spatial symmetries, the decay is sensitive to the small components of the ^{8}Li wave function that are poorly constrained and and sensitive to the details of the diagonalization of the 1^+; 1 states. The agreement with data for the remaining $A = 8$ transitions involving the ground states of ^{8}Li and its mirror ^{8}B to the first 2^+ state in ^{8}Be may be further impacted by an experimental difficulty arising from the possible presence of a 2^+ intruder state [18], as discussed for the VMC results. These transitions will require further scrutiny both from the theoretical and experimental points of view to determine the precise cause of the discrepancies.

4.1.2 Two-Body GT Transition Densities

To better understand the dynamics underlying the difference in the $3N$ fit, I computed two-body transition densities. These are defined such that,

$$M^{\rm 2b}_{\rm GT} = 4\,\pi \int_0^\infty dr\, r^2\, \rho^{\rm 2b}(r)\,, \tag{4.5}$$

where r is the interparticle distance, and 2b = N2LO-Δ, N2LO-OPE, and N2LO-CT. Practically speaking, such densities are computed by evaluating the matrix element [25, 26],

$$\rho(r) = \frac{1}{4\pi r^2} \langle \Psi_f | \sum_{i<j} \mathcal{O}_{ij} \delta(r - |\mathbf{r}_i - \mathbf{r}_j|) | \Psi_i \rangle\,, \tag{4.6}$$

where $\mathcal{O}_{ij}$ is any generic two-body operator, $\mathbf{r}_i$ indicates the coordinate of particle i, and the matrix element between the nuclear wave functions $\Psi_{i/f}$ is computed via Monte Carlo integration. These calculations, originally presented in Ref. [13],

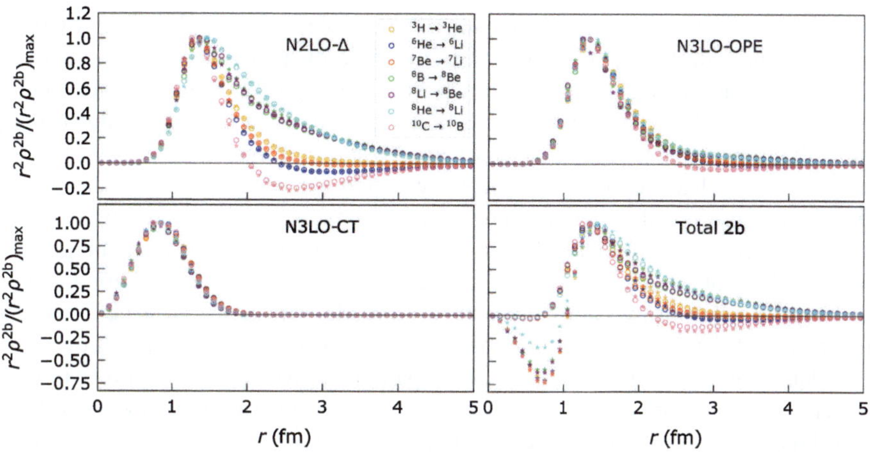

Fig. 4.3 Two-body transition density defined as in Eq. 4.5 for selected transitions (represented with different colors) obtained with the NV2+3-Ia (open circles) and NV2+3-Ia* models (filled stars). Figure reproduced with permission from Ref. [13]

represent the first such two-body axial transition densities computed in a microscopic framework.

Figure 4.3 displays the two-body transitions densities scaled such that the point on the curve with the largest magnitude peaks at 1. From the total two-body transition density, one sees that there is a larger negative contribution at short-ranges in Model Ia* relative to model Ia coming from the N3LO-CT term. Interestingly, the N3LO-CT panel shows that the shape of this two-body transition density is the same across all transitions and models. Thus, for a given transition, models Ia and Ia* differ only by an overall scaling factor. This behavior is well understood in light of the fact that the N3LO-CT term has its strength governed by one LEC, $z_0 \propto c_D$. Thus, how one fits the $3N$ force inherently influences this term.

The universal shape may be understood from the fact that the GT β-decay converts an $ST = 10$ ($ST = 01$) pair to a pair with $ST = 01$ ($ST = 10$) [27]. At short distances, relative P-waves are suppressed and $ST = 10$ and $ST = 01$ pairs dominate. The fact that the $ST = 10$ and $ST = 01$ pair densities are similar in shape to each other and exhibit a universal short-range behavior across different nuclei [28] is reflected in the short-range GT transition densities. The $ST = 11$ and $ST = 00$ pair densities, on the other hand, display no such universality. Since P-waves become more important with larger interparticle distance, the presence of these pairs spoils the universal behavior in the tails.

The formation of $ST = 11$ pairs is particularly sensitive to the tensor potential and, thus, the NN interaction, while the formation of $ST = 00$ pairs are generally disfavored due to the strong repulsion in that channel [26, 29]. The increasing importance of $ST = 11$ pairs with large spacing combined with their sensitivity to the tensor term in the NV2 leads to a tail that is consistent between the two models, but dependent on the particular transition. Because of an interplay between two operator structures in the N2LO-Δ and N3LO-OPE terms that cancel one another out, these contributions need not be strictly positive or negative. In fact, if one focuses on the $^{10}\mathrm{C} \rightarrow {}^{10}B$ decay, a node is clearly present in the long-range total two-body transition density. With the enhanced negative contribution from the N3LO-CT in model Ia*, this leads to an overall negative correction when two-body currents are included. Therefore, the conclusion one draws from this systematic study of different χEFT interactions is that correlations account for the bulk of the quenching in the light nuclei that were studied. Two-body currents can lead to a reduction in M_{GT}; however, the sign of two-body currents has a model sensitivities arising due to the fit of the $3N$ force and a structural dependence arising from the impact of tensor correlations on the formation of $ST = 11$ pairs in a given system. In order to fully understand the nature of the g_A quenching problem, further study in the two-body and three-body sectors is necessary to understand these sensitivities across models with the same NN potential that differ in the $3N$ force. Further, using the NV2+3 models in methods amenable for the study of heavier systems could help to determine if these conclusions change when one goes beyond the p-shell.

4.2 B(GT) for Charge Exchange Reactions

Along with comparing GT matrix elements extracted directly from β-decay lifetimes, I also benchmarked them against values of GT strengths B(GT) extracted from charge exchange reactions [30–32]. The strength function is connected to the matrix element via the relation,

$$B(\mathrm{GT}) = \frac{|M_{\mathrm{GT}}|^2}{2J_i + 1} . \tag{4.7}$$

Although the charge exchange reaction involves a strongly interacting probe, the same operator structure that mediates the GT transition gives rise to this reaction. One may decompose the charge-exchange cross section $d\sigma/d\Omega$ into different components of angular momentum ΔL. By fitting the cross section and extracting the $\Delta L = 0$ component, one can obtain the B(GT) from the relation,

$$\frac{d\sigma}{d\Omega}(q = 0)|_{\Delta L=0} = \hat{\sigma}_{\mathrm{GT}} B(\mathrm{GT}) , \tag{4.8}$$

where $\hat{\sigma}_{\mathrm{GT}}$ is the unit cross section, a proportionality factor depending on kinematics and the details of the nucleon-nucleus interaction used to analyze the charge exchange cross section. By comparing the B(GT) extracted in this way with QMC calculations, it is possible to learn about the wave functions of exotic nuclei using β-decays that are difficult to probe directly.

To compare with $B(GT)$ from excited state decays in the proton-rich nucleus ^{11}N to the ^{11}C ground state that were extracted with (p, n) reactions, I computed these quantities in VMC. My calculations and shell model evaluations were compared to the experimental data in Ref. [33] and both approaches compared favorably with the measurements. In particular, the VMC values of $B(\text{GT})[1/2^- \rightarrow \text{g.s.}] = 0.2050(7)$ and $B(\text{GT})[3/2^- \rightarrow \text{g.s.}] = 0.180(1)$ agreed within error with the experimental values of $0.18(1)^{\text{stat}}(3)^{\text{sys}}$ and $0.18(1)^{\text{stat}}(4)^{\text{sys}}$, respectively. It is also interesting to compare the $B(\text{GT})$ coming from the mirror transitions of ^{11}Be to ^{11}B. If continuum couplings effects are significant for the decays of these exotic nuclei, one should expect a large difference in the decay strengths of the mirror nuclear states. There was, however, a discrepancy in different charge exchange extractions coming from $(d,{}^2\text{He})$ [34] and $(t,{}^3\text{He})$ [35] reactions. In particular, the $(t,{}^3\text{He})$ indicated a small change in the mirror transition strengths, while the $(d,{}^2\text{He})$ was consistent with a larger difference and more significant continuum coupling effects. When evaluated with VMC, it was found that there was a only a small difference in the mirror transitions, with values of $B(\text{GT})[1/2^- \rightarrow \text{g.s.}] = 0.2012(4)$ and $B(\text{GT})[3/2^- \rightarrow \text{g.s.}] = 0.175(1)$. These are in good agreement with the extracted values from $(t,{}^3\text{He})$ of $B(\text{GT})[1/2^- \rightarrow \text{g.s.}] = 0.23(5)$ and $B(\text{GT})[3/2^- \rightarrow \text{g.s.}] = 0.17(5)$. The values from $(d,{}^2\text{He})$ of $B(\text{GT})[1/2^- \rightarrow \text{g.s.}] \approx 0.34$ and $B(\text{GT})[3/2^- \rightarrow \text{g.s.}] \approx 0.33$ did not agree with the VMC evaluation, but are consistent with the values obtained in an unquenched shell model evaluation. It thus was concluded that there was no significant difference in the single-particle properties of the low-lying states in these near-dripline mirror nuclei and that they behaved consistent with the expectation for the p-shell.

4.3 The ^{6}He β-Decay Spectrum for BSM Physics Searches

To this point, I have discussed the validation of the NV2+3 model with measured β-decay rates; however, this framework is also useful for predicting SM processes relevant to BSM physics searches. In particular, β-decay spectra provide a sensitive probe of charge-changing currents that are not included in the SM. The interference of BSM currents with the V-A currents of the SM would induce a distinctive dependence in the β-decay spectrum on the ratio of the electron mass to the outgoing electron energy, m_e/E_e. This dependence is the so-called "Fierz interference term" [36], usually denoted by b. The first constraint $-0.018 < b < 0.052$ came from direct neutron measurements [37, 38] and measurements of super-allowed $0^+ \rightarrow 0^+$ β-decays constrain the Fierz interference term induced by scalar currents to $b < 3.3 \cdot 10^{-3}$ at the 90% confidence level [1]. Measuring purely GT β-decay spectra allows one to constrain b from tensor and pseudoscalar currents. On-going measurements are targeting a 0.1% uncertainty for the purely GT ^{6}He β-decay spectrum [39–41]. This would provide a rather clear signature of b, as the shape of the spectrum for this specific decay would be [36],

$$\frac{d\Gamma}{dE_e} = \frac{d\Gamma_0}{dE_e}\left[1 + b\frac{m_e}{E_e}\right], \tag{4.9}$$

where $d\Gamma_0/dE_e$ is the SM spectrum. With the proposed experimental precision, deviations of the measurement from the SM prediction would indicate new physics up to the 10 TeV scale [42]; however, this is relies on theoretical calculations of the SM spectrum with comparable uncertainties. The picture is further complicated, as it is also possible to generate a non-zero value of b by accounting for recoil corrections that come from the small momentum transferred to the nucleus in the decay [43, 44].

Because of the need for accurate theory calculations, it is crucial to approach the problem in a many-body framework that is highly accurate for light nuclei. The first such calculation of the ^{6}He $\rightarrow$ ^{6}Li β-decay spectrum was performed in Ref. [8], using the VMC wave functions and the AV18+UIX Hamiltonian. Ref. [45] performed a modern calculation of this observable using the NCSM approach with chiral potentials and LO one-body transition operators [45] to obtain $b = -1.52(18) \times 10^{-3}$ coming from SM recoil corrections. Most recently, a QMC study of both SM and BSM nuclear matrix elements for the ^{6}He $\rightarrow$ ^{6}Li decay was carried out in Ref. [46] with the aim of providing a theoretical uncertainty on b below the permille sensitivity aim of on-going experiments. For this work, I computed the relevant matrix elements with the NV2+3 interaction and consistent electroweak currents including two-body physics. Using these calculations and the standard multipole analysis, my collaborators and I published the first study of this quantity to incorporate subleading weak current contributions and we provided an estimate of the NV2+3 model uncertainty, which reduced the theoretical error to below 0.1%.

4.3.1 The β-Decay Rate

The differential rate for the β-decay of ^{6}He is derived in Sect. B.2. Here, I present the final result obtained using the standard multipole analysis, which is given by,

$$\begin{aligned} d\Gamma = &(2\pi)\delta(E_i - E_f - E_\nu - E_e)\frac{1}{2J_i+1}G_F^2 V_{ud}^2 \frac{4\pi}{x+1} \\ &\Big[(1 + \hat{\mathbf{k}}_\nu \cdot \hat{\mathbf{k}}_e)|C_1(q)|^2 + (1 - \hat{\mathbf{k}}_\nu \cdot \hat{\mathbf{k}}_e + 2(\hat{\mathbf{k}}_\nu \cdot \hat{\mathbf{q}})(\hat{\mathbf{q}} \cdot \hat{\mathbf{k}}_e))|L_1(q;A)|^2 \\ &-\hat{\mathbf{q}} \cdot (\hat{\mathbf{k}}_\nu + \hat{\mathbf{k}}_e)2\mathrm{Re}(L_1(q;A)C_1^*(q;a))\Big] + \Big[(1 - (\hat{\mathbf{k}}_\nu \cdot \hat{\mathbf{q}})(\hat{\mathbf{q}} \cdot \hat{\mathbf{k}}_e))(|M_1(q;V)|^2 \\ &+|E_1(q;A)|^2) + \hat{\mathbf{q}} \cdot (\hat{\mathbf{k}}_\nu - \hat{\mathbf{k}}_e)2\mathrm{Re}(M_1(q;V)E_1^*(q;A))\Big], \end{aligned} \tag{4.10}$$

where $V_{ud} = 0.97373(31)$ is the element of the Cabibbo-Kobayashi-Maskawa (CKM) matrix which describes the strength with which the up quark couples to the down quark via the weak interaction. From the definitions in Eqs. B.1 through B.4, we may write the four multipoles entering the rate as,

$$C_1(q;A) = \frac{i}{\sqrt{4\pi}} \langle {}^6\mathrm{Li}, 10|\rho_+^\dagger(q\hat{\mathbf{z}};A)|{}^6\mathrm{He}, 00\rangle \tag{4.11}$$

$$L_1(q;A) = \frac{i}{\sqrt{4\pi}} \langle {}^6\mathrm{Li}, 10|\hat{\mathbf{z}}\cdot\mathbf{j}_+^\dagger(q\hat{\mathbf{z}};A)|{}^6\mathrm{He}, 00\rangle \tag{4.12}$$

$$E_1(q;A) = -\frac{i}{\sqrt{2\pi}} \langle {}^6\mathrm{Li}, 10|\hat{\mathbf{z}}\cdot\mathbf{j}_+^\dagger(q\hat{\mathbf{x}};A)|{}^6\mathrm{He}, 00\rangle \tag{4.13}$$

$$M_1(q;V) = -\frac{1}{\sqrt{2\pi}} \langle {}^6\mathrm{Li}, 10|\hat{\mathbf{y}}\cdot\mathbf{j}_+^\dagger(q\hat{\mathbf{x}};V)|{}^6\mathrm{He}, 00\rangle\,; \tag{4.14}$$

however, noting that the momentum transfer is limited by the reaction Q-value of ≈ 3.5 MeV $\ll m_\pi$, it is useful to expand the currents in powers of $q/m_\pi = qr_\pi$. Because the multipoles are defined in terms of spherical Bessel functions and their derivatives—which are either purely even or odd functions of q—we can write the following expansions,

$$C_1(q;A) = -i\frac{qr_\pi}{3}\left(C_1^{(1)}(A) - \frac{(qr_\pi)^2}{10}C_1^{(3)}(A) + \mathcal{O}\left[(qr_\pi)^4\right]\right), \tag{4.15}$$

$$L_1(q;A) = -\frac{i}{3}\left(L_1^{(0)}(A) - \frac{(qr_\pi)^2}{10}L_1^{(2)}(A) + \mathcal{O}\left[(qr_\pi)^4\right]\right), \tag{4.16}$$

$$M_1(q;V) = -i\frac{qr_\pi}{3}\left(M_1^{(1)}(V) - \frac{(qr_\pi)^2}{10}M_1^{(3)}(V) + \mathcal{O}\left[(qr_\pi)^4\right]\right), \tag{4.17}$$

$$E_1(q;A) = -\frac{i}{3}\left(E_1^{(0)}(A) - \frac{(qr_\pi)^2}{10}E_1^{(2)}(A) + \mathcal{O}\left[(qr_\pi)^4\right]\right). \tag{4.18}$$

The utility of this expansion, combined with the fact that the NV2+3 electroweak currents themselves admit an expansion in m_π/Λ_χ, is that the coefficients in Eqs. 4.11 through 4.13 can be used to express the spectrum with a precision up to some desired order q/Λ_χ. To achieve the below a 0.1% accuracy required for on-going experiments, one may neglect terms of $\mathcal{O}(q^3/\Lambda_\chi^3)$. Working to this order, summing over all spins, and integrating Eq. 4.10 over E_ν and the angle between the outgoing leptons, the rate becomes,

$$\frac{d\Gamma}{d\varepsilon} = (1+\Delta_R^V)(1+\delta_R(Z,\varepsilon))\frac{G_F^2 W_0^5 V_{ud}^2}{2\pi^3}\sqrt{1-\frac{\mu_e^2}{\varepsilon^2}}\,\varepsilon^2(1-\varepsilon)^2 F_0(Z,\varepsilon)L_0(Z,\varepsilon)S(Z,\varepsilon)R_N(\varepsilon)$$
$$\frac{4\pi}{2J_i+1}\frac{1}{9}\Bigg\{3\left|L_1^{(0)}\right|^2\left[1+\alpha Z W_0 R\left(\frac{2}{35}-\frac{233}{630}\frac{\alpha Z}{W_0 R}-\frac{1}{70}\frac{\mu_e^2}{\varepsilon}-\frac{4}{7}\varepsilon\right)\right]$$
$$+2W_0 r_\pi\left[\left(1-2\varepsilon+\frac{\mu_e^2}{\varepsilon}\right)\mathrm{Re}(E_1^{(0)}M_1^{(1)*})-\left(1-\frac{\mu_e^2}{\varepsilon}\right)\mathrm{Re}(L_1^{(0)}C_1^{(1)*})\right]$$
$$+\frac{(W_0 r_\pi)^2}{3}\left[\left(3-4\varepsilon(1-\varepsilon)-\mu_e^2\frac{2+\varepsilon}{\varepsilon}\right)|C_1^{(1)}|^2-\frac{3}{5}\left(1-\frac{\mu_e^2}{\varepsilon}(2-\varepsilon)\right)\mathrm{Re}\left(L_1^{(0)}L_1^{(2)*}\right)\right.$$
$$\left.+\left(3-10\varepsilon(1-\varepsilon)+\mu_e^2\frac{4-7\varepsilon}{\varepsilon}\right)\left(\left|M_1^{(1)}\right|^2-\frac{1}{5}\mathrm{Re}\left(E_1^{(0)}E_1^{(2)}\right)\right)\right]$$
$$-\frac{4}{7}\frac{\alpha Z W_0 r_\pi^2}{R}(1-\varepsilon)\left(\frac{E_1^{(0)}E_1^{(2)}}{2}-L_1^{(0)}L_1^{(2)}\right)\Bigg\}, \tag{4.19}$$

where the scaled variables $E_e = W_0\varepsilon$ and $m_e = W_0\mu_e$ have been introduced, with $W_0 = M_i - M_f = 4.016$ MeV in the case of the ^{6}He $\rightarrow$ ^{6}Li transition. M_i and M_f are the masses of the initial and final state nuclei. The effects of nuclear recoil are retained to leading order in W_0/M_f. Further, nuclear recoil makes it so that the electron endpoint energy shifts from $E_e = W_0$ to $E_e = W_0 - \frac{W_0^2-m_e^2}{2M_f}$.

In addition to the multipoles, the above expression also depends on a number of radiative corrections that are non-negligible at the precision with which Eq. 4.19 is derived. Rather than derive electromagnetic corrections in χEFT, Ref. [46] followed the standard approach to incorporating radiative corrections, as summarized in Ref. [47]. In the future, consistently including radiative corrections to β-decays in χEFT will be important and work along these lines is being pursued by the community [48]. In the approach adopted for this work, the well-known correction from the Fermi function is present and accounts for the Coulomb distortion of the outgoing electron wave function. It is given by $F_0(Z,\varepsilon)$, defined as,

$$F_0(Z,\varepsilon) = 4(2|\mathbf{p}_e|R)^{2(\gamma_0-1)}\frac{|\Gamma(\gamma_0+iy)|^2}{|\Gamma(2\gamma_0+1)|^2}e^{\pi y}, \quad \gamma_0=\sqrt{1-(\alpha Z)^2}, \quad y=\frac{\alpha Z}{|v_e|}, \tag{4.20}$$

where $R=\sqrt{5/3\langle r_{\rm ch}^2\rangle}$, and $\langle r_{\rm ch}^2\rangle$ is the charge radius of ^{6}Li, $\sqrt{\langle r_{\rm ch}^2\rangle} = 2.5890(390)$ fm [49]. Z is the charge of the daughter nucleus and the electron velocity $v_e = |\mathbf{p}_e|/E_e$. The so called "inner" and "outer" radiative corrections, labelled Δ_R^V and $\delta_R(Z,\varepsilon)$, respectively, contribute to the half-life at the level of few percent. These and other higher order corrections contained in the functions $L_0(Z,\varepsilon)$, $S(Z,\varepsilon)$, $R_N(\varepsilon)$, and in the explicit $\mathcal{O}(\alpha)$ and $\mathcal{O}(\alpha^2)$ dependence of Eq. 4.19, are discussed further in Appendix C. It may also be noted that the leading contribution to SM rate is

$$\frac{d\Gamma_0}{d\varepsilon} = \frac{G_F^2 W_0^5 V_{ud}^2}{2\pi^3}\sqrt{1-\frac{\mu_e^2}{\varepsilon^2}}\,\varepsilon^2(1-\varepsilon)^2\frac{4\pi}{3}\left|L_1^{(0)}\right|^2 F_0(Z,\varepsilon)L_0(Z,\varepsilon)S(Z,\varepsilon)R_N(\varepsilon), \tag{4.21}$$

which still includes some radiative corrections. By comparing Eq. 4.19 to Eq. 4.21, it is possible to connect with the definition of b in Eq. 4.9.

4.3.2 Procedure to Obtain QMC Matrix Elements

To determine the coefficients in the polynomial expansions of the multipoles in Eqs. 4.15 through 4.18, there is a natural strategy; namely, I evaluated a given multipole at several values of q and fit the curve to an even or odd polynomial as required. In the study of Ref. [46], the grid adopted was six points spaced by 0.05 fm^{-1} between $q = 0$ fm^{-1} and $q = 0.25$ fm^{-1}. The effect of doubling the number of points in the interval, as well as to adding or removing a point, were tested and I found that the results were not significantly altered by these changes. The fits were performed to the polynomial expressions in Eqs. 4.15 to 4.18, but to ensure convergence, the coefficients were also extracted when including higher order terms in the fit. The coefficients obtained in the odd polynomials were unchanged by including the $\mathcal{O}((qr_\pi)^5)$ term and had the same statistical uncertainty. For the even polynomials, higher order terms did not alter the central values outside within statistical uncertainty; however, including each successive higher-order coefficient reduced the uncertainty by orders of magnitude. Because of this, it was determined that going to higher orders in the even polynomials could lead to an underestimate of the uncertainty; therefore, the expansion was truncated at $\mathcal{O}((qr_\pi)^2)$ to prevent overfitting, though this could lead to a conservative result for the estimated statistical error.

To have a realistic assessment of the theoretical uncertainty, the fit of the expansion coefficients was performed to multipoles obtained with four sets of chiral potentials and currents; namely, the NV2+3-Ia*, NV2+3-IIa*, NV2+3-IIb*, and NV2+3-Ia. Results were obtained with both the VMC and GFMC methods. Figure 4.4 illustrates the procedure by showing the calculated VMC multipoles and the associated curves of best fit for all four models. Table 4.3 displays the VMC coefficients for all models, where the error denotes the statistical error coming from the least squares fitting procedure. Two-body currents have a minor effect on the L_1 and E_1 multipoles, leading to a shift in $L_1^{(0)}$, $E_1^{(0)}$, $L_1^{(2)}$, $E_1^{(2)}$ of at most $\sim$2%, in the case of model Ia. This is consistent with the few percent correction to M_{GT}, as these quantities are inherently related. Two-body corrections enter at NLO in the vector charge and axial current operators. As such, two-body currents should be important for M_1 and C_1. Indeed, it was observed that $M_1^{(1)}$ receives an $\sim$8% correction from two-body currents for all the models considered here. In VMC, the corrections to $C_1^{(1)}$ are about 30% to 40%.

The expansion coefficients for the NV2+3-Ia*, NV2+3-IIa*, NV2+3-IIb*, and NV2+3-Ia models are presented in Table 4.4. From Fig. 4.4 and Table 4.3, there is clearly an uncertainty due to the choice of model. To account for this, four sets of average multipoles were obtained: An average between two models with the same cutoff, using the same fitting procedure for the $3N$ LECs, and fit to different ranges of

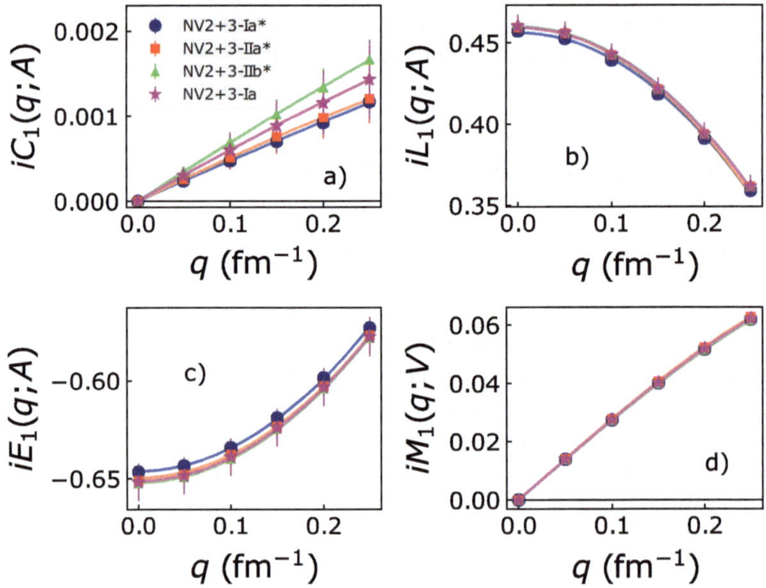

Fig. 4.4 VMC multipoles for the NV2+3-Ia* (blue circles), NV2+3-IIa* (orange squares), NV2+3-IIb* (green triangles) and NV2+3-Ia (magenta stars) models. The curves of best fit for each case are shown in the same color as the multipoles. Statistical errors from the Monte Carlo are included on each point but are too small to be visible in the figure. Figure reproduced with permission from Ref. [46]

Table 4.3 Expansion coefficients of the VMC multipole operators, including only one-body currents (1b) or one- and two-body currents (2b). The four columns denotes four different NV2+3 models and the error is associated with the least squares fitting procedure

	Model Ia*		Model IIa*		Model IIb*		Model Ia	
	1b	1b+2b	1b	1b+2b	1b	1b+2b	1b	1b+2b
$L_1^{(0)}$	1.3578(3)	1.3607(4)	1.3662(2)	1.3742(3)	1.3717(2)	1.3777(1)	1.3641(3)	1.3986(3)
$E_1^{(0)}$	1.9255(4)	1.9299(5)	1.9355(2)	1.9471(3)	1.9470(3)	1.9555(1)	1.9333(4)	1.9824(4)
$M_1^{(1)}$	−0.5487(1)	−0.5860(2)	−0.5510(1)	−0.5952(2)	−0.5550(1)	−0.60111(4)	−0.5521(1)	−0.5908(2)
$C_1^{(1)}$	−0.0182(3)	−0.0269(3)	−0.0217(3)	−0.0311(3)	−0.0151(3)	−0.0257(3)	−0.0128(3)	−0.0218(3)
$L_1^{(2)}$	23.87(5)	23.93(6)	23.51(3)	23.51(4)	23.94(3)	24.07(1)	23.64(5)	24.12(5)
$E_1^{(2)}$	17.86(7)	17.99(8)	17.35(5)	17.49(6)	18.07(5)	18.22(2)	17.45(7)	17.80(7)

NN scattering data (NV2+3-Ia* and NV2+3-IIa*); an average between two models fit to the same range of *NN* scattering data, using the same fitting procedure for the $3N$ LECs, and different cutoffs (NV2+3-IIa* and NV2+3-IIb*); an average between two models with consistent NV2 and cutoff but different NV3 interactions (NV2+3-Ia* and NV2+3-Ia); and an average of all four models. These average multipoles were fit to the functional forms of Eqs. (4.15)–(4.18) with the two model calculations at each q providing the upper and lower theoretical error bar. The statistical uncertainty of

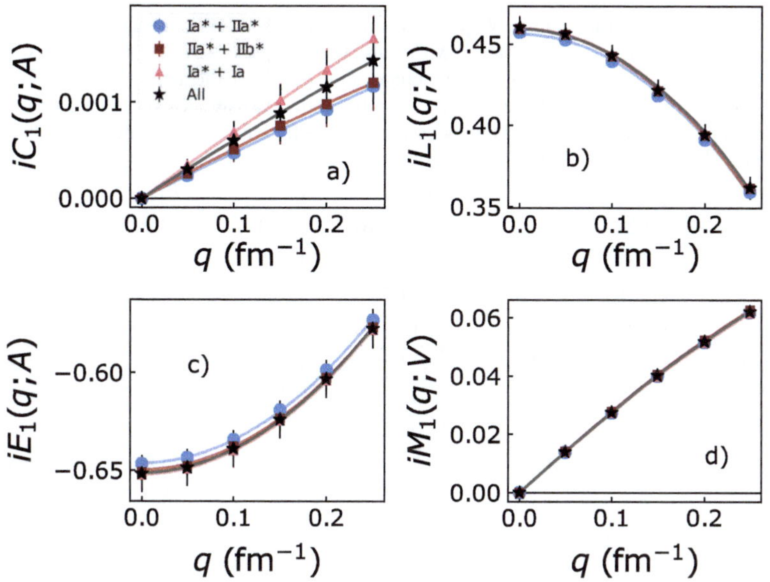

Fig. 4.5 Average VMC multipoles for the four NV2+3 models under study. Averages were obtained for NV2+3 potentials with the same cutoff (blue circles), the same energy range of *NN* scattering data used to fit the interaction (red squares), the same *NN* but different $3N$ force (pink triangles), and for all models (black stars). The curves of best fit for each case are shown in the same color as the average multipoles. Details on how the error bars are obtained in each case are provided in the text. Figure reproduced with permission from Ref. [46]

these averaged fits provides the estimated uncertainty on the expansion coefficients due to the energy range of the fit (ϵ_E), the choice of cutoff (ϵ_C), and the three-body force fitting procedure (ϵ_{3N}).

For the fourth case described above, a similar procedure was followed; however, when assigning the error for the multipoles at each value of q, the uncertainty was not taken as simply the spread of the model calculations. Instead, the estimated cutoff, energy range, and $3N$ uncertainties on each point were summed in quadrature to obtain the total error. The expansion coefficients obtained from this fit are the "Average" results in Table 4.4 and their statistical uncertainties are cited as ϵ_{tot}. Figure 4.5 shows the averaged multipoles and curves of best fit for these three cases.

As can be seen in Table 4.4, the coefficient $E_1^{(2)}$ has the largest model uncertainty at 15%. The next largest uncertainties are on the coefficients $C_1^{(1)}$ at 11.8% and $L_1^{(2)}$ at 7.8%. The remaining uncertainties are $\lesssim 1\%$ for $L_1^{(0)}$, $E_1^{(0)}$, and $M_1^{(1)}$. The $3N$ fit is the main source of uncertainty for the coefficient of the even polynomials, while for $C_1^{(1)}$ and $M_1^{(1)}$, the uncertainties from this source and the others are comparable. This may be appreciated as due to the connection between c_D in the $3N$ force and z_0 in the N3LO-CT axial current, as was discussed at length in Sect. 4.1. It is the energy range of scattering data used to fit the interaction that provides the largest uncertainty on $M_1^{(1)}$, which represents the dominant subleading correction to Eq. 4.21.

Table 4.4 Summary of the values of the expansion coefficients for the charge (C_1), longitudinal (L_1), electric (E_1), and magnetic (M_1) VMC multipoles for the NV2+3 models under study. The percent error due to the cutoff $\epsilon_C(\%)$, energy range of the fit $\epsilon_E(\%)$, the three-body force fit $\epsilon_{3N}(\%)$, and the total error $\epsilon_{\rm tot}(\%)$ are also presented. Details of how the average expansion coefficients are obtained are provided in the text. The uncertainty on coefficients for individual models is fitting error only. The error is < 0.001 unless otherwise noted

Model	$C_1^{(1)}$	$L_1^{(0)}$	$L_1^{(2)}$	$E_1^{(0)}$	$E_1^{(2)}$	$M_1^{(1)}$
Ia*	−0.027	1.361	23.93(6)	1.923	17.99(8)	−0.586
IIa*	−0.031	1.374	23.51(4)	1.947	17.49(6)	−0.595
IIb*	−0.026	1.378	24.07(1)	1.955	18.22(2)	−0.601
Ia	−0.022	1.399	24.12(5)	1.982	17.80(7)	−0.591
Average	−0.026	1.378	23.869	1.954	17.847	−0.593
$\epsilon_E(\%)$	5.4	0.3	3.2	0.3	5.6	0.6
$\epsilon_C(\%)$	6.9	0.01	0.3	0.07	0.8	0.3
$\epsilon_{3N}(\%)$	7.2	0.8	7.0	0.8	13.9	0.3
$\epsilon_{\rm tot}(\%)$	11.8	0.8	7.8	0.9	15.0	0.7

Another way that one might obtain the average expansion parameters with an uncertainty would be to instead use the results of Table 4.3. I computed the average value of the expansion coefficients using the four models and this differed from the values extracted in the adopted procedure by less than one percent. One could estimate the uncertainty due to the energy range used to fit *NN* interaction from the difference between the NV2+3-Ia* and NV2+3-IIa* results, the uncertainty due to the cutoff from the difference between the NV2+3-IIa* and NV2+3-IIb* results, and the uncertainty due to the three-body force from the difference between the NV2+3-Ia* and NV2+3-Ia results. Following this procedure, the error on the leading order expansion coefficients increase by factors of 1.5 to 1.8. The error on the dominant contributions would be $\lesssim 1.4\%$ in this approach. For the coefficients $L_1^{(2)}$ and $E_1^{(2)}$, this scheme reduces the uncertainty by a factors of 2.6 and 3.0, respectively; however, the adopted approach in Ref. [46] is more reasonable, as each matrix element should have its own model uncertainty that then propagates to the expansion coefficients in the fitting procedure.

In addition to a calculation of the multipoles using VMC, a GFMC propagation was also performed for the relevant matrix elements. The GFMC matrix elements were fit following the same procedure used for the VMC calculations. The results of these fits are presented in Table 4.5 for models Ia*, IIa*, and Ia. The GFMC evolution reduces the leading coefficients, $L_1^{(0)}$ and $E_1^{(0)}$, by ~3-4%, which, as will be discussed later, results in better agreement with the experimental half-life. Propagating models Ia* and Ia changes the remaining expansion coefficients at the level of a few percent, while for model IIa*, terms of higher order in qr_π changed more significantly after the propagation. In Ref. [46], this was explained to be related to the the system size as a function of τ during the GFMC propagation. The point proton radii were computed in GFMC and, for model IIa*, the system size grows much more rapidly

Table 4.5 Expansion coefficients of the GFMC multipoles compared the VMC results. The calculations was performed using three different Norfolk models. The error denotes the fitting error only

Model	Method	$C_1^{(1)}$	$L_1^{(0)}$	$L_1^{(2)}$	$E_1^{(0)}$	$E_1^{(2)}$	$M_1^{(1)}$
Ia*	VMC	−0.027	1.361	23.93(6)	1.930	17.99(8)	−0.586
	GFMC	−0.047	1.308	23.98(6)	1.856	19.71(9)	−0.559
IIa*	VMC	−0.031	1.374	23.51(4)	1.947	17.49(6)	−0.595
	GFMC	−0.045(2)	1.341(7)	25(1)	1.901(9)	19(2)	−0.573(4)
Ia	VMC	−0.022	1.399	24.12(5)	1.982	17.80(7)	−0.591
	GFMC	−0.045	1.360	23.64(6)	1.929	17.77(9)	−0.575
Average	VMC	−0.026(3)	1.38(1)	24(2)	1.95(2)	18(3)	−0.593(4)
	GFMC	−0.046(2)	1.34(2)	24(3)	1.90(3)	19(5)	−0.568(9)

in τ than models Ia and Ia*. This behavior is due to the proximity of the ^{6}He(0^+; 1) (^{6}Li(1^+; 0)) ground state energy in GFMC to the $\alpha + 2n$ ($\alpha + d$) breakup threshold. Because of terms going like $e^{-i\mathbf{q}\cdot\mathbf{r}_i}$ in the current operators, the monotonic increase of the system size will impact the convergence of the IIa* matrix elements needed to determine the multipoles. To account for this in the GFMC extrapolation with model IIa*, we adopt a procedure used for systems near threshold and broad resonances [22]. While the system sizes grow with τ, the ground state energies of ^{6}Li and ^{6}He drop rapidly and stabilize near $\tau \approx 0.1$ MeV^{-1}. The stability of the energy indicates that spurious contamination has been removed from the wave functions at that point. Under this assumption, the values of the matrix elements were extracted by performing a linear fit to the matrix element in the interval $\tau = [0.1\ \text{MeV}^{-1}, 0.3\ \text{MeV}^{-1}]$ and extrapolating back to 0.1 MeV^{-1}. The systematic error of this procedure was determined by also averaging in the intervals $\tau = [0.08\ \text{MeV}^{-1}, 0.3\ \text{MeV}^{-1}]$ and $\tau = [0.12\ \text{MeV}^{-1}, 0.3\ \text{MeV}^{-1}]$ to get a conservative estimate for the error bar.

For the GFMC fits, the matrix elements have uncertainty assigned from the energy range and three-body force following the same procedure as was done for the VMC. The additional systematic uncertainty from the model IIa* extrapolation was summed in quadrature with the other sources of uncertainty to obtain a total error bar on the matrix elements. In the GFMC average ressult, the coefficients $L_1^{(0)}$, $E_1^{(0)}$, $C_1^{(1)}$, and $M_1^{(1)}$ have uncertainties ranging from ∼1.4% to ∼5.2%. The coefficients $L_1^{(2)}$ and $E_1^{(2)}$ have uncertainties of ∼12% and ∼26%, respectively. In the expression for the rate, the quadratic coefficients appear suppressed by $(qr_\pi)^2$ and thus do not contribute as strongly to the uncertainty on the spectrum as the coefficients with errors of order 1%.

One may note that only three models were used in the GFMC propagation. This choice was made in Ref. [46] because of the small cutoff uncertainty. To estimate the impact of including model IIb*, the sum of the matrix elements was re-weighted in the average so that model IIa* is counted twice. Averaging under this assumption provides coefficients and errors that are consistent with the average results obtained when omitting model IIb*. Thus, neglecting model IIb* should not significantly alter the conclusions.

4.3.3 SM Spectrum Results

Figure 4.6 shows the deviation of the SM spectrum using VMC and GFMC with the associated theory uncertainties. The theoretical uncertainties originate from the error estimation on the QMC coefficients, as well as from the uncertainty on fundamental constants in the rate. In principle, there is a truncation uncertainty associated with the chiral and polynomial expansions. The fist term left out in each multipole should contribute less than the model uncertainties, though this is based on the power counting of the charge and current. In principle, robust uncertainty quantification should be carried out in the future, though no such procedures have not yet been developed for electroweak transition matrix elements. There is vigorous work along these lines in the nuclear many-body community, but such an uncertainty quantification was beyond the scope of the work in Ref. [46]. The error estimate is showed in the bottom panel of Fig. 4.6.

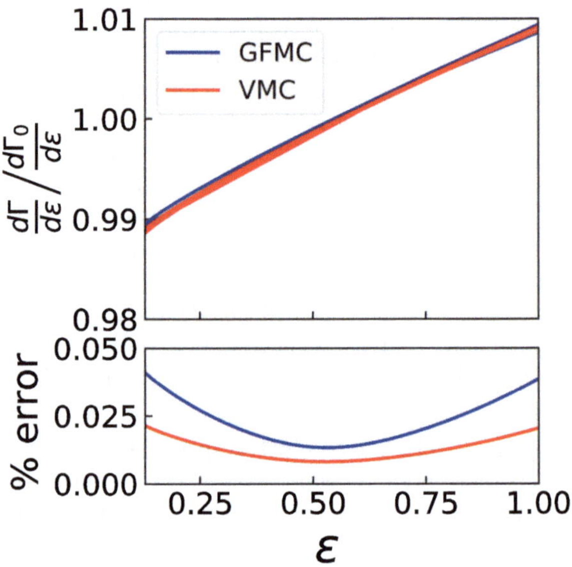

Fig. 4.6 Deviation of the ^{6}He β-decay spectrum from the expression truncated at leading order in the multipole expansion, given in Eq. (4.21), and percentage error on the ratio. The blue curves use GFMC matrix elements in Table 4.5, while the red lines come from the VMC averages in Table 4.4. In the top panel, the width of the curves denotes the theoretical error. Figure reproduced with permission from Ref. [46]

While the VMC and GFMC provide different integrated rates, they have relatively similar shapes. Even with the enhanced systematic uncertainty from extrapolating the mutlipoles with the NV2+3-IIa*, the GFMC spectrum has an error well below the permille required by on-going experimental analyses. It is also possible to use the GFMC matrix elements to study the corrections to the spectrum. From Ref. [46], we have that,

$$\frac{d\Gamma}{d\varepsilon}\bigg|_{\mathrm{GFMC}} = \frac{d\Gamma_0}{d\varepsilon}\Bigg\{1+\left[\left(1-2\varepsilon+\frac{\mu_e^2}{\varepsilon}\right)(-1.15\pm 0.02)+\left(1-\frac{\mu_e^2}{\varepsilon}\right)(6.6\pm 0.7)\cdot 10^{-2}\right]\cdot 10^{-2}$$
$$-\left[\left(1-\frac{\mu_e^2}{\varepsilon}(2-\varepsilon)\right)(0.99\pm 0.12)+\left(3-10\varepsilon(1-\varepsilon)+\mu_e^2\frac{4-7\varepsilon}{\varepsilon}\right)(0.35\pm 0.10)\right]\cdot 10^{-3}$$
$$+(4.0\pm 1.8)\cdot 10^{-4}\,(1-\varepsilon)\Bigg\}. \tag{4.22}$$

where the terms in the first line appear at NLO in the multipole expansion, and are proportional to $M_1^{(1)}$ and $C_1^{(1)}$, respectively. The terms in the second line appear at N2LO, the first proportional to $L_1^{(2)}$ and the second to a combination of $|M_1^{(1)}|^2$ and $E_1^{(0)}E_1^{(2)}$. Lastly, the term in the third line is an electromagnetic correction proportional to subleading multipoles. The dominant correction to the spectrum is from $M_1^{(1)}$, while the leading source of error on the correction actually comes from the quadratic multipoles due to the large systematic uncertainty on these terms in the GFMC propagation. From the correction proportional to μ_e/ε, one can extract b from the SM due to recoil corrections. The SM correction $b = -1.47(3)\times 10^{-3}$ obtained with QMC not only agrees with the previous NCSM evaluation [45], but also significantly reduces the uncertainty by including explicit two-body physics.

4.3.4 *Cross-Validation of the QMC Results*

To verify that the computational approach and fitting procedure obtained reasonable results, there were a few validations performed in Ref. [46]. First, given the connection between the $q \to 0$ behavior $L_1(q)$ and $E_1(q)$ and the total rate, one may validate the $L_1^{(0)}$ and $E_1^{(0)}$ multipoles by comparing our calculation with experimental half-life $\tau_{1/2} = 807.25 \pm 0.16 \pm 0.11$ ms [50, 51]. From the VMC matrix elements in Table 4.4, one obtains

$$\tau_{1/2}\big|_{\mathrm{VMC}} = \frac{\log 2}{\Gamma} = \frac{1}{V_{ud}^2 g_A^2}\,(1175\pm 17)\ \mathrm{ms} = (762 + 11 \pm 2)\ \mathrm{ms} \tag{4.23}$$

where $V_{ud} = 0.97370 \pm 0.00031$ and $g_A = 1.2754 \pm 0.0013$ [52]. The first error is due to the nuclear matrix elements, while the second arises from g_A. The error from V_{ud} is negligible. Eq. (4.23) deviates by about 5% from the observed value. While small, the discrepancy between the VMC calculation and the experiment is not accounted for by the uncertainty range in Eq. (4.23). Using the more accurate GFMC matrix elements, the results is

$$\tau_{1/2}\big|_{\text{GFMC}} = \frac{1}{V_{ud}^2 g_A^2}\,(1246 \pm 37)\ \text{ms} = (808 \pm 24 \pm 2)\ \text{ms}, \tag{4.24}$$

which has a 3% error and is in excellent agreement with the observed half-life. The half-life is dominated by $L_1^{(0)}$, $E_1^{(0)}$, and the Fermi function in Eq. 4.20. The next most important correction comes from inner and outer radiative corrections, Δ_R^V and $\delta_R(Z, \varepsilon)$, which together shorten the half-life by about 4%. The adopted value of $\Delta_R^V = 2.467(22) \cdot 10^{-2}$ came from the dispersive evaluation of Refs. [53–55]. Higher multipoles only contribute 0.1% of the uncertainty on the half-life. The absolute spectra that integrate to give these rates are plotted in Fig. 4.7.

The magnetic multipole $M_1(q)$ may also be cross-checked against data. Using the conserved vector current hypothesis, $M_1^{(1)}$ can be expressed in terms of the electromagnetic transition $\Gamma_{M_1}(^6\text{Li}(0^+) \to ^6\text{Li}(1^+)\gamma)$,

$$\left|M_1^{(1)}\right| = \frac{3m_\pi}{\sqrt{4\pi}}\sqrt{\frac{\Gamma_{M_1}}{\alpha E_\gamma^3}} = 0.582(6) \tag{4.25}$$

where the uncertainty is determined by the width, $\Gamma_{M1} = 8.19(17)$ eV [49, 56]. Using the average VMC value listed in Table 4.4, the calculations agrees with the experimental value within 1%, while individual models are consistent within 3% when including two body currents. The GFMC average is a bit lower, but still only 2% away from Eq. (4.25), and compatible within $\sim 1\sigma$. The LO GFMC value also agrees with the NCSM calculation of Ref. [45], which, including only the one-body piece of the current, finds $|M_1^{(1)}| = 0.565$.

Equation (4.25) is valid up to isospin breaking corrections. The level of isospin breaking expected for $M_1^{(1)}$ was estimated in Ref. [46] by computing both

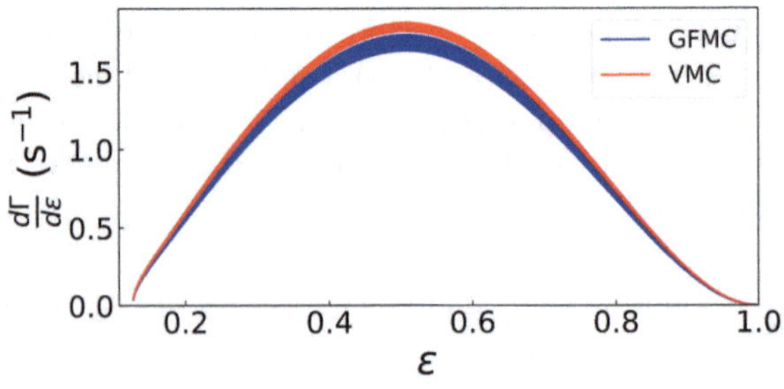

Fig. 4.7 The absolute ^{6}He β-decay spectrum from the expression truncated at leading order in the multipole expansion, given in Eq. (4.21). The blue band is obtained with the GFMC matrix elements in Table 4.5, while the red band is from the VMC averages in Table 4.4. The width of the curves denote the theoretical error

the $^6\mathrm{He}(0^+; 1) \rightarrow ^6\mathrm{Li}(1^+; 0)$ transition and the electromagnetic transition $^6\mathrm{Li}(0^+; 1) \rightarrow ^6\mathrm{Li}(1^+; 0)$ with NV2+3-IIb wave functions at LO in χEFT. Using VMC wave functions, the electromagnetic transition had a value 0.553 and the weak transition a value of 0.554. Agreement between the two M_1 calculations is thus achieved at the ~0.1% level. After GFMC propagation, in the electromagnetic case, $|M_1^{(1)}|_{\mathrm{EM}} = 0.532$ and, for the weak transition, $|M_1^{(1)}| = 0.538$, showing a 0.9% difference. This analysis is performed with one nuclear interaction model and thus is not accounting for any model uncertainty.

The change in the size of the isospin breaking between VMC and GFMC can be understood based on how the wave functions are generated in each method. In VMC, the variational parameters in the $^6\mathrm{He}(0^+; 1)$ and $^6\mathrm{Li}(0^+; 1)$ wave functions are minimized separately. The explicit isospin symmetry breaking terms in the potential result in different variational parameters in the two wave functions; however, isospin symmetry breaking correlations are not turned on in the VMC wave functions. When the trial states are acted on with the imaginary time propagator, isospin breaking correlations are introduced. This, in turn, increases the disagreement for the electromagnetic and weak M_1 transitions. Because the systems only differ by changing the isospin of one nucleon, the effect of this symmetry breaking is small and at the level of the experimental uncertainty in Eq. (4.25). Because isospin breaking corrections are smaller than the experimental error, it is thus possible to replace the theoretical value of $M_1^{(1)}$ with that extracted from $^6\mathrm{Li}(0^+; 1) \rightarrow ^6\mathrm{Li}(1^+; 0)$ to further reduce the error on M_1 to the 1% level, as was done in Ref. [46].

Finally, the verification of the $C_1^{(1)}$ expansion coefficient may only be carried out by comparing with experiment, as no data exist to benchmark this quantity. As seen in Eq. A.99, the axial charge receives a $\mathcal{O}(1/m_N)$ recoil contribution from the coupling of the axial current to the nucleon, and a contribution from the induced pseudoscalar form factor, proportional to the energy of the electron and neutrino $E_e + E_\nu = W_0$. The matrix element of the recoil component, however, vanishes for transitions between the dominant S state components of the wave function [43], making it particularly sensitive to the mixing of different $LS[n]$ components in wave function and to two-body currents. The QMC calculations are indeed dominated by the pseudoscalar contribution. Accounting also for two-body effects, we may decompose C_1 as (see Sect. A.3),

$$C_1^{(1)} = C_{\mathrm{recoil}}^{(1)} + C_{\mathrm{pseudo}}^{(1)} + C_{2-\mathrm{body}}^{(1)}, \tag{4.26}$$

The earliest—and until recently, the only—theoretical determination of the axial charge contribution to ^{6}He β-decay was performed by Calaprice [43]. A direct comparison can be made by observing [57]

$$C_{\mathrm{recoil}}^{(1)} = -\sqrt{\frac{3(2J_i + 1)}{4\pi}} \frac{d(0)}{2r_\pi M} \tag{4.27}$$

where $d(q^2)$ is the induced tensor form factor. Using wave functions tuned to reproduce the experimental energy levels, Calaprice obtained $d(0) = 2.4$ [43], which can be converted into $C^{(1)}_{\rm recoil} = -0.0144$. This value is in the same ballpark of the VMC calculation, which however has an opposite sign. After GFMC evolution, the recoil contribution is reduced and qualitatively agrees with the NCSM result of Ref. [45].

In Ref. [46], this discrepancy was investigated further leveraging the fact that,

$$C^{(1)}_{\rm recoil} = -\sqrt{\frac{3}{4\pi}}\frac{g_A}{2r_\pi m_N}\left(\mathcal{M}_{\sigma L} + m_N\mathcal{M}_{\sigma rp}\right), \tag{4.28}$$

where $\mathcal{M}_{\sigma L}$ and $\mathcal{M}_{\sigma rp}$ are one-body matrix elements defined in Ref. [44]. Calaprice neglected the latter matrix element, assuming that the $M_{\sigma L}$ term was dominant. The $\mathcal{M}_{\sigma L}$ contribution was evaluated with the NV2+3-Ia* interaction in VMC and GFMC calculations. The parameters β_{LS}, determining the coupling of the p-shell nucleons to different LS-channels (see Eq. 3.6), for the NV2+3-Ia* are $\beta_{00} = 0.931$ and $\beta_{11} = -0.364$ for ^{6}He. For ^{6}Li, the components are given by $\beta_{01} = 0.967$, $\beta_{10} = 0.182$ and $\beta_{21} = 0.176$. These couplings are not dissimilar from the shell model LS couplings in Ref. [43], though the D wave component was found to be smaller ($\beta_{21} = -0.03$). For the VMC, the value obtained is,

$$C^{(1)}_{\rm recoil}\Big|_{\sigma L} = 0.013, \tag{4.29}$$

with statistical uncertainties < 0.001. There is an approximate linear dependence on β_{21}, which would shift the value to 0.012 for $\beta_{21} = -0.03$. Following Ref. [10], the "one-pion-exchange-like" correlation operators in the VMC wave functions may be turned off to mimic the shell model. Doing so increases the matrix element by 10%, from 0.013 to 0.014. Thus, while the magnitude of the matrix element agrees very well with Ref. [43], there is still a discrepancy in the sign.

GFMC propagation further decreaes the contribution to

$$C^{(1)}_{\rm recoil}\Big|_{\sigma L} = 0.009. \tag{4.30}$$

This, of course, is only within one model and does not account for any uncertainty from using different NV2+3 models; however, in Ref. [46], the following conclusions were reached: (i) the magnitude of $\mathcal{M}_{\sigma L}$ agrees well with shell model calculations, but is quenched by correlations, (ii) the contribution of $\mathcal{M}_{\sigma rp}$ is non-negligible in the case of ^{6}He, and, as expected in Refs. [43, 58–60], and (iii) two-body currents make a substantial contribution to $C_1(q)$.

Finally, the induced pseudoscalar contribution was not considered in Ref. [43]. In agreement with the NCSM result of Ref. [45], this contribution dominates $C^{(1)}_1$ in the QMC calculations.

4.3.5 BSM Contribution to the Spectrum

In Ref. [46], the effects of BSM physics on the spectrum were analyzed using an effective field theory approach [61, 62]. Assuming that BSM physics contributes a scale much larger than the one associated with electroweak physics, it may be incorporated in an effective field theory that adds to the SM the most general set of gauge-invariant effective operators expressed in terms of SM fields and organized according to their canonical dimension. This approach is referred to as "Standard Model effective field theory" or SMEFT.

The standard multipole expansion may be expanded in terms of BSM currents induced by SMEFT operators, as was done in Ref. [46] and, earlier, in Ref. [63]. The lowest order psuedoscalar (P) and tensor (T) contributions in SMEFT at the nucleon level are given by the following one-body operators,

$$\rho_{P,a}^{(-4)} = \frac{g_A B}{m_\pi^2 + \mathbf{q}^2} \frac{\tau_{1,a}}{2} \boldsymbol{\sigma}_1 \cdot \mathbf{q} (2\pi)^3 \delta(\mathbf{k}_1 - \mathbf{q}) + 1 \leftrightarrow 2, \tag{4.31}$$

$$j_{T,a}^{ij(-3)} = g_T \varepsilon^{ijk} \frac{\tau_{1,a}}{2} \sigma_1^k (2\pi)^3 \delta(\mathbf{k}_1 - \mathbf{q}) + 1 \leftrightarrow 2, \tag{4.32}$$

$$j_{T,a}^{i0(-2)} = i \frac{g_T'}{2m_N} q^i \frac{\tau_{1,a}}{2} (2\pi)^3 \delta(\mathbf{k}_1 - \mathbf{q}) + 1 \leftrightarrow 2\,, \tag{4.33}$$

where g_T is the isovector tensor charge, $g_T = 0.989 \pm 0.033$ [64], and $B = m_\pi^2/(m_u + m_d) {\sim} 2.8$ GeV. The coupling g_T' is, to this point, unknown; however, it should be a number of order 1. The superscript in parentheses indicates the power counting in χEFT. Defining,

$$\rho_P(\mathbf{q}) = \int d^3\mathbf{x}\, e^{i\mathbf{q}\cdot\mathbf{x}} \rho_P(\mathbf{x}) \tag{4.34}$$

$$\mathbf{j}_T^{(\prime)}(\mathbf{q}) = \int d^3\mathbf{x}\, e^{i\mathbf{q}\cdot\mathbf{x}} \mathbf{j}_T^{(\prime)}(\mathbf{x})\,, \tag{4.35}$$

it is possible to derive the multipoles in a similar fashion as for the vector and axial currents. It is also possible to include sterile neutrinos, which have the same multipoles from Fourier transforming vector and axial current contributions. By the selection rules for GT β-decay with ΔJ=1, only the multipoles $C_1(q, P)$, $L_1(q, T)$, $E_1(q, T)$ and $M_1(q, T')$ will contribute. These multipoles have a low q expansion just like Eqs. (4.15)–(4.18). The coefficient, $M_1^{(0)}(T')$ is zero and, given the power counting of $\mathbf{j}^{(\prime)}$, higher order coefficients will contribute below the level of precision obtained in the SM spectrum. Thus, only the multipoles $C_1(q, P)$, $L_1(q, T)$, and $E_1(q, T)$ need to be considered. It turns out that, to first order [46],

$$L_1^{(0)}(T) = \frac{1}{\sqrt{2}} E_1^{(0)}(T) = -\frac{2g_T}{g_A} L_1^{(0)}(A), \quad (4.36)$$

$$C_1^{(1)}(P) = -\frac{B}{m_\pi} L_1^{(0)}(A)\,, \quad (4.37)$$

which made it is possible to incorporate the first order signals of new physics using the QMC calculations. Because sterile neutrinos introduce terms depending on the same standard multipoles, their effect is also easily incorporated from the QMC results.

With tensor and pseudoscalar couplings corresponding to BSM physics at scales 4 and 8 TeV, respectively, there is an observable deviation from the recoil-corrected SM spectrum obtained in QMC. In Fig. 4.8, the results of including the first order tensor and pseudoscalar are show by the red and green curves, respectively. Indeed, the interference of the axial and tensor current generates a distortion with the m_e/E_e dependence of the Fierz interference term. In Appendix D, I discuss the physical interpretation of this shape. Because tensor operator is chirality-flipping, in the ultrarelativistic limit when $E_e \gg m_e$, flipping the electron chirality is the same as flipping its helicity. Conservation of helicity should mean that it is not possible to generate any distortion from interference between tensor and axial contributions. Indeed, the tensor and axial would sum coherently at very large electron energies with no interference. At lower energies, however, it is possible to generate interference between the two contributions and hence we can qualitatively understand the energy dependence. As discussed in Appendix D, the interference may be constructive or destructive and would depend on the sign of the tensor coupling ϵ_T, which

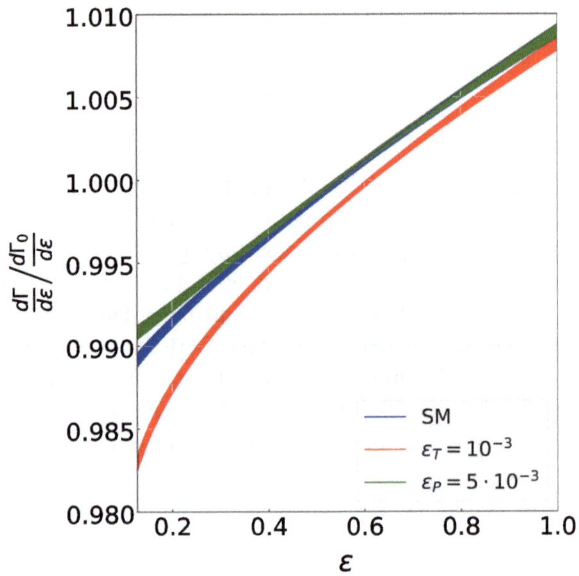

Fig. 4.8 The deviation of ^{6}He β-decay spectrum from Eq. 4.21 (blue) compared with the deviations generated in SMEFT by tensor (red) and pseudoscalar (green) contributions to the charge changing weak current. Model uncertainties are represented by the width of the bands. Figure reproduced with permission from Ref. [46]

need not be positive or negative. Thus, Fig. 4.8 represents just one possibility for the shape of the spectrum including a particular choice for ϵ_T. The peusdoscalar interference, instead, is generated with an additional factor $(1 - \varepsilon)$, which is why it vanishes at large ε.

While present constraints from pion decays [65] make the observation of a pseudoscalar contribution at this scale unlikely, a tensor coupling of this size is not ruled out [66–68]. Thus, on-going experiments will either observe or constrain new physics with the ^{6}He β-decay spectrum when analyzing the data using the GFMC calculations using the NV2+3 interaction in Ref. [46].

It is also possible to observe the effect of including a sterile neutrino of mass $m_{\nu_4} =$ 1 MeV that has strength U_{e4} to mix with the electron type neutrino. Measurements of the ^{20}F and ^{144}Pr β-decay spectra constrain the mixing angle in this mass range—assuming that the mixing is the only interaction between the sterile neutrino and the SM—to be $|U_{e4}|^2 \sim 2 \cdot 10^{-3}$ [69–74]. Figure 4.9 shows that a sterile neutrino at this mass with a mixing $|U_{e4}|^2 \sim 10^{-3}$ and assuming only SM currents would generate a characteristic kink in the spectrum. In the right panel, a scenario with a quark level tensor coupling constant $\widetilde{\varepsilon}_T = 3 \cdot 10^{-2}$ is shown, which would indicate physics at the 1 TeV scale. This coupling, assuming $|U_{e4}|^2 = 0.5 \cdot 10^{-3}$ results in a significant distortion to the spectrum. Again, comparing to the QMC results, present experimental efforts will be able to place constraints on sterile neutrinos with masses $m_{\nu 4} \approx 1$ MeV.

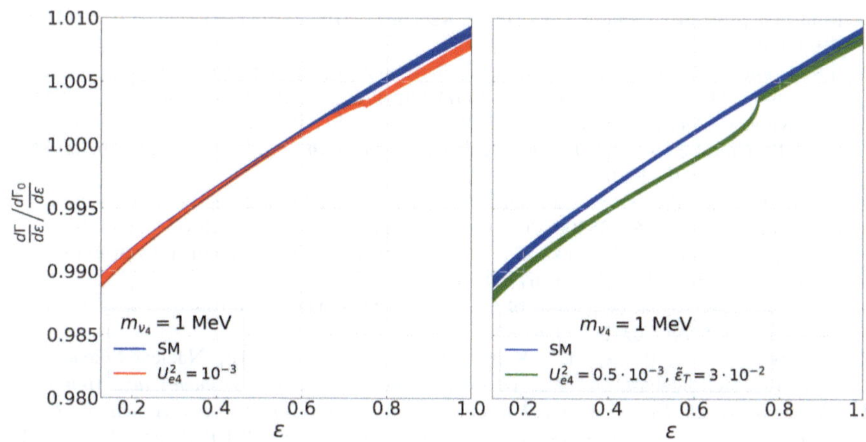

Fig. 4.9 (LEFT) The deviation o ^{6}He β-decay spectrum from Eq. 4.21 (blue) compared with the deviation generated in SMEFT by a sterile neutrino with mass $m_{\nu_4} = 1$ MeV and mixing $|U_{e4}|^2 = 10^{-3}$. (RIGHT) The SM spectrum compared with the spectrum distorted by a sterile neutrino of mass $m_{\nu_4} = 1$ MeV with mixing Model uncertainties are represented by the width of the bands. Figure reproduced with permission from Ref. [46]

References

1. J.C. Hardy, I.S. Towner, Superallowed $0^+ \rightarrow 0^+$ nuclear β decays: 2020 critical survey, with implications for V_{ud} and CKM unitarity. Phys. Rev. C **102**(4), 045501 (2020). https://doi.org/10.1103/PhysRevC.102.045501
2. W.T. Chou, E.K. Warburton, B. Alex Brown, Gamow-Teller beta-decay rates for A <= 18 nuclei. Phys. Rev. **C47**, 163–177 (1993). https://doi.org/10.1103/PhysRevC.47.163
3. G.F. Bertsch, I. Hamamoto, Gamow-teller strength at high excitations. Phys. Rev. C **26**, 1323–1326 (1982). https://doi.org/10.1103/PhysRevC.26.1323
4. M.G.E. Brand, K. Allaart, W.H. Dickhoff, Conserving rpa and the response of 48 ca. Phys. Lett. B **214**, 483–489 (1988). https://doi.org/10.1016/0370-2693(88)90104-9
5. M.G.E. Brand, K. Allaart, W.H. Dickhoff, Nuclear response beyond mean field theory. Nucl. Phys. A **509**, 1–38 (1990). https://doi.org/10.1016/0375-9474(90)90374-U
6. M. Ichimura, H. Sakai, T. Wakasa, Spin-isospin responses via (p, n) and (n, p) reactions. Prog. Part. Nucl. Phys. **56**, 446–531 (2006). https://doi.org/10.1016/j.ppnp.2005.09.001
7. M. Ichimura, Spin-isospin responses of nuclei: gamow-teller response functions and pionic response functions in the quasi-elastic scattering region. Nucl. Phys. B, Proc. Suppl. **159**, 197–202 (2006). https://doi.org/10.1016/j.nuclphysbps.2006.08.038
8. R. Schiavilla, R.B. Wiringa, Weak transitions in A = 6 and A = 7 nuclei. Phys. Rev. C **65**, 054302 (2002). https://doi.org/10.1103/PhysRevC.65.054302
9. M. Pervin, S.C. Pieper, R.B. Wiringa, Quantum monte carlo calculations of electroweak transition matrix elements in a = 6,7 nuclei. Phys. Rev. C **76**, 064319 (2007). https://doi.org/10.1103/PhysRevC.76.064319
10. S. Pastore, A. Baroni, J. Carlson, S. Gandolfi, Steven C. Pieper, R. Schiavilla, R.B. Wiringa, Quantum Monte Carlo calculations of weak transitions in $A = 6 - 10$ nuclei. Phys. Rev. **C97**(2), 022501 (2018). https://doi.org/10.1103/PhysRevC.97.022501
11. P. Gysbers et al., Discrepancy between experimental and theoretical β-decay rates resolved from first principles. Nature Phys. **15**(5), 428–431 (2019). https://doi.org/10.1038/s41567-019-0450-7
12. S.K. Bogner, R.J. Furnstahl, R.J. Perry, Similarity renormalization group for nucleon-nucleon interactions. Phys. Rev. C **75**, 061001 (2007). https://doi.org/10.1103/PhysRevC.75.061001
13. G.B. King, L. Andreoli, S. Pastore, M. Piarulli, R. Schiavilla, R.B. Wiringa, J. Carlson, S. Gandolfi, Chiral effective field theory calculations of weak transitions in light nuclei. Phys. Rev. C **102**, 025501 (2020)
14. A. Knecht et al., Precision measurement of the he-6 half-life and the weak axial current in nuclei. Phys. Rev. C **86**, 035506 (2012). https://doi.org/10.1103/PhysRevC.86.035506
15. T. Suzuki, R. Fujimoto, T. Otsuka, Gamow-teller transitions and magnetic properties of nuclei and shell evolution. Phys. Rev. C **67**, 044302 (Apr2003)
16. E.K. Warburton, R-matrix analysis of the $\beta^{\mp}$-delayed alpha spectra from the decay of ^{8}Li and ^{8}b. Phys. Rev. C **33**, 303–313 (Jan1986)
17. B. Pritychenko, E. Běták, M.A. Kellett, B. Singh, J. Totans, The Nuclear Science References (NSR) database and Web Retrieval System. *Nuclear Instruments and Methods in Physics Research Section A: Accelerators, Spectrometers, Detectors and Associated Equipment*, 640(1):213-218, 2011. ISSN 0168-9002. https://doi.org/10.1016/j.nima.2011.03.018. http://www.sciencedirect.com/science/article/pii/S0168900211005584
18. M.T. Burkey et al., Improved limit on tensor currents in the weak interaction from li8 β decay. Phys. Rev. Lett. **128**(20), 202502 (2022). https://doi.org/10.1103/PhysRevLett.128.202502
19. A. Baroni, L. Girlanda, S. Pastore, R. Schiavilla, M. Viviani, Nuclear axial currents in chiral effective field theory. Phys. Rev. C **93**(1), 015501 (2016). https://doi.org/10.1103/PhysRevC.93.049902, DOI: https://doi.org/10.1103/PhysRevC.93.015501 10.1103/PhysRevC.95.059901. [Erratum: Phys. Rev. C 95, no.5,059901(2017)]
20. A. Baroni et al., Local chiral interactions, the tritium gamow-teller matrix element, and the three-nucleon contact term. Phys. Rev. C **98**(4), 044003 (2018). https://doi.org/10.1103/PhysRevC.98.044003

21. R.B. Wiringa, S.C. Pieper, J. Carlson, V.R. Pandharipande, Quantum Monte Carlo calculations of A = 8 nuclei. Phys. Rev. C **62**, 014001 (2000). https://doi.org/10.1103/PhysRevC.62.014001
22. S. Pastore, R.B. Wiringa, Steven C. Pieper, R. Schiavilla, Quantum Monte Carlo calculations of electromagnetic transitions in ^{8}Be with meson-exchange currents derived from chiral effective field theory. Phys. Rev. **C90**(2), 024321 (2014). https://doi.org/10.1103/PhysRevC.90.024321
23. V.M. Datar et al., Electromagnetic transition from the 4+ to 2+ resonance in be8 measured via the radiative capture in he4+he4. Phys. Rev. Lett. **111**(6), 062502 (2013). https://doi.org/10.1103/PhysRevLett.111.062502
24. J.C. Hardy, I.S. Towner, Superallowed $0^+ \rightarrow 0^+$ nuclear β decays: 2014 critical survey, with precise results for V_{ud} and CKM unitarity. Phys. Rev. C **91**(2), 025501 (2015). https://doi.org/10.1103/PhysRevC.91.025501
25. R.B. Wiringa, R. Schiavilla, S.C. Pieper, J. Carlson, Nucleon and nucleon-pair momentum distributions in $A \leq 12$ nuclei. Phys. Rev. C **89**(2), 024305 (2014). https://doi.org/10.1103/PhysRevC.89.024305
26. M. Piarulli, S. Pastore, R.B. Wiringa, S. Brusilow, R. Lim, Densities and momentum distributions in a≤12 nuclei from chiral effective field theory interactions. Phys. Rev. C **107**(1), 014314 (2023). https://doi.org/10.1103/PhysRevC.107.014314
27. R. Schiavilla et al., Weak capture of protons by protons. Phys. Rev. C **58**, 1263 (1998). https://doi.org/10.1103/PhysRevC.58.1263
28. J.L. Forest, V.R. Pandharipande, Steven C. Pieper, Robert B. Wiringa, R. Schiavilla, and A. Arriaga. Femtometer toroidal structures in nuclei. Phys. Rev. **C54**, 646–667 (1996). https://doi.org/10.1103/PhysRevC.54.646
29. R.B. Wiringa, Pair counting, pion-exchange forces, and the structure of light nuclei. Phys. Rev. C **73**, 034317 (2006). https://doi.org/10.1103/PhysRevC.73.034317
30. T.N. Taddeucci, C.A. Goulding, T.A. Carey, R.C. Byrd, C.D. Goodman, C. Gaarde, J. Larsen, D. Horen, J. Rapaport, E. Sugarbaker, The (p, n) reaction as a probe of beta decay strength. Nucl. Phys. A **469**, 125–172 (1987). https://doi.org/10.1016/0375-9474(87)90089-3
31. R. Meharchand et al., Probing configuration mixing in be-12 with gamow-teller transition strengths. Phys. Rev. Lett. **108**, 122501 (2012). https://doi.org/10.1103/PhysRevLett.108.122501
32. S. Giraud et al., β+ Gamow-Teller Strengths from Unstable O14 via the (d, He2) Reaction in Inverse Kinematics gamow-teller strengths from unstable o14 via the (d, he2) reaction in inverse kinematics. Phys. Rev. Lett. **130**(23), 232301 (2023). https://doi.org/10.1103/PhysRevLett.130.232301
33. J. Schmitt et al., Probing spin-isospin excitations in proton-rich nuclei via the c11(p, n)n11 reaction. Phys. Rev. C **106**(5), 054323 (2022). https://doi.org/10.1103/PhysRevC.106.054323
34. T. Ohnishi et al., Study of spin-isospin excitations in 11 be via the (d, 2 he) reaction at 270 mev. Nucl. Phys. A **687**, 38–43 (2001). https://doi.org/10.1016/S0375-9474(01)00598-X
35. I. Daito et al., Gamow-teller strengths from (t, 3 he) charge-exchange reactions on light nuclei. Phys. Lett. B **418**, 27–33 (1998). https://doi.org/10.1016/S0370-2693(97)01393-2
36. J.D. Jackson, S.B. Treiman, H.W. Wyld, Possible tests of time reversal invariance in beta decay. Phys. Rev. **106**, 517–521 (1957). https://doi.org/10.1103/PhysRev.106.517
37. H. Saul, C. Roick, H. Abele, H. Mest, M. Klopf, A. Petukhov, T. Soldner, X. Wang, D. Werder, B. Märkisch, Limit on the fierz interference term b from a measurement of the beta asymmetry in neutron decay. Phys. Rev. Lett. **125**(11), 112501 (2020). https://doi.org/10.1103/PhysRevLett.125.112501
38. X. Sun et al., Improved limits on fierz interference using asymmetry measurements from the ultracold neutron asymmetry (ucna) experiment. Phys. Rev. C **101**(3), 035503 (2020). https://doi.org/10.1103/PhysRevC.101.035503
39. A. Garcia et al., http://faculty.washington.edu/agarcia3/Chirality-flipping/
40. W. Byron et al., First observation of cyclotron radiation from MeV-scale e$^\pm$ following nuclear beta decay. 9 (2022)
41. M. Kanafani, X. Fléchard, O. Naviliat-Cuncic, G.D. Chung, S. Leblond, E. Liénard, X. Mougeot, G. Quéméner, A. Simancas Di Filippo, J.-C. Thomas, Precision measurements in

the beta decay of ^{6}He. EPJ Web Conf. **282**, 01010 (2023). https://doi.org/10.1051/epjconf/202328201010
42. V. Cirigliano, A. Garcia, D. Gazit, O. Naviliat-Cuncic, G. Savard, A. Young, Precision Beta Decay as a Probe of New Physics. 7 (2019)
43. F.P. Calaprice, Second class interactions and the electron-neutrino correlation in nuclear beta decay. Phys. Rev. C **12**(6), 2016–2021 (1975). ISSN 05562813. https://doi.org/10.1103/PhysRevC.12.2016
44. B.R. Holstein, Recoil effects in allowed beta decay: the elementary particle approach. Rev. Mod. Phys. **46**, 789 (1974). https://doi.org/10.1103/RevModPhys.46.789. [Erratum: Rev. Mod. Phys. 48, 673-673 (1976)]
45. A. Glick-Magid, C. Forssén, D. Gazda, D. Gazit, P. Gysbers, P. Navrátil, Nuclear ab initio calculations of 6he β-decay for beyond the standard model studies. Phys. Lett. B **832**, 137259 (2022). https://doi.org/10.1016/j.physletb.2022.137259
46. G.B. King, A. Baroni, V. Cirigliano, S. Gandolfi, L. Hayen, E. Mereghetti, S. Pastore, M. Piarulli, Ab initio calculation of the β-decay spectrum of he6. Phys. Rev. C **107**(1), 015503 (2023). https://doi.org/10.1103/PhysRevC.107.015503
47. L. Hayen, N. Severijns, K. Bodek, D. Rozpedzik, X. Mougeot, High precision analytical description of the allowed $$ spectrum shape. Rev. Modern Phys. **90**(1), 015008 (2018). ISSN 0034-6861. https://doi.org/10.1103/RevModPhys.90.015008
48. V. Cirigliano, W. Dekens, E. Mereghetti, O. Tomalak, Effective field theory for radiative corrections to charged-current processes: vector coupling. Phys. Rev. D **108**(5), 053003 (2023). https://doi.org/10.1103/PhysRevD.108.053003
49. D.R. Tilley, C.M. Cheves, J.L. Godwin, G.M. Hale, H.M. Hofmann, J.H. Kelley, C.G. Sheu, H.R. Weller, Energy levels of light nuclei a=5, a=6, a=7. Nucl. Phys. A **708**, 3–163 (2002). https://doi.org/10.1016/S0375-9474(02)00597-3
50. M. Kanafani, X. Fléchard, O. Naviliat-Cuncic, G.D. Chung, S. Leblond, E. Liénard, X. Mougeot, G. Quéméner, A. Simancas Di Filippo, J.C. Thomas, High precision measurement of the ^{6}he half-life. Phys. Rev. C **106**(4), 045502 (2022). https://doi.org/10.1103/PhysRevC.106.045502
51. A. Knecht et al., Precision measurement of the 6he half-life and the weak axial current in nuclei. Phys. Rev. Lett. **108**, 122502 (2012). https://doi.org/10.1103/PhysRevLett.108.122502
52. M. Tanabashi, K. Hagiwara, K. Hikasa, K. Nakamura, Y. Sumino, F. Takahashi, J. Tanaka, K. Agashe, G. Aielli, C. Amsler, M. Antonelli, D.M. Asner, H. Baer, Sw. Banerjee, R.M. Barnett, T. Basaglia, C.W. Bauer, J.J. Beatty, V.I. Belousov, J. Beringer, S. Bethke, A. Bettini, H. Bichsel, O. Biebel, K.M. Black, E. Blucher, O. Buchmuller, V. Burkert, M.A. Bychkov, R.N. Cahn, M. Carena, A. Ceccucci, A. Cerri, D. Chakraborty, M.-C. Chen, R.S. Chivukula, G. Cowan, O. Dahl, G. D'Ambrosio, T. Damour, D. de Florian, A. de Gouvêa, T. DeGrand, P. de Jong, G. Dissertori, B.A. Dobrescu, M. D'Onofrio, M. Doser, M. Drees, H.K. Dreiner, D.A. Dwyer, P. Eerola, S. Eidelman, J. Ellis, J. Erler, V.V. Ezhela, W. Fetscher, B.D. Fields, R. Firestone, B. Foster, A. Freitas, H. Gallagher, L. Garren, H.-J. Gerber, G. Gerbier, T. Gershon, Y. Gershtein, T. Gherghetta, A.A. Godizov, M. Goodman, C. Grab, A.V. Gritsan, C. Grojean, D.E. Groom, M. Grünewald, A. Gurtu, T. Gutsche, H.E. Haber, C. Hanhart, S. Hashimoto, Y. Hayato, K.G. Hayes, A. Hebecker, S. Heinemeyer, B. Heltsley, J.J. Hernández-Rey, J. Hisano, A. Höcker, J. Holder, A. Holtkamp, T. Hyodo, K.D. Irwin, K.F. Johnson, M. Kado, M. Karliner, U.F. Katz, S.R. Klein, E. Klempt, R.V. Kowalewski, F. Krauss, M. Kreps, B. Krusche, Y.V. Kuyanov, Y. Kwon, O. Lahav, J. Laiho, J. Lesgourgues, A. Liddle, Z. Ligeti, C.-J. Lin, C. Lippmann, T.M. Liss, L. Littenberg, K.S. Lugovsky, S.B. Lugovsky, A. Lusiani, Y. Makida, F. Maltoni, T. Mannel, A.V. Manohar, W.J. Marciano, A.D. Martin, A. Masoni, J. Matthews, U.-G. Meißner, D. Milstead, R.E. Mitchell, K. Mönig, P. Molaro, F. Moortgat, M. Moskovic, H. Murayama, M. Narain, P. Nason, S. Navas, M. Neubert, P. Nevski, Y. Nir, K.A. Olive, S. Pagan Griso, J. Parsons, C. Patrignani, J.A. Peacock, M. Pennington, S.T. Petcov, V.A. Petrov, E. Pianori, A. Piepke, A. Pomarol, A. Quadt, J. Rademacker, G. Raffelt, B.N. Ratcliff, P. Richardson, A. Ringwald, S. Roesler, S. Rolli, A. Romaniouk, L.J. Rosenberg, J.L. Rosner, G. Rybka, R. A. Ryutin, C.T. Sachrajda, Y. Sakai, G.P. Salam, S. Sarkar, F. Sauli, O. Schneider,

K. Scholberg, A.J. Schwartz, D. Scott, V. Sharma, S.R. Sharpe, T. Shutt, M. Silari, T. Sjöstrand, P. Skands, T. Skwarnicki, J.G. Smith, G.F. Smoot, S. Spanier, H. Spieler, C. Spiering, A. Stahl, S.L. Stone, T. Sumiyoshi, M.J. Syphers, K. Terashi, J. Terning, U. Thoma, R.S. Thorne, L. Tiator, M. Titov, N.P. Tkachenko, N. A. Törnqvist, D.R. Tovey, G. Valencia, R. Van de Water, N. Varelas, G. Venanzoni, L. Verde, M.G. Vincter, P. Vogel, A. Vogt, S.P. Wakely, W. Walkowiak, C.W. Walter, D. Wands, D.R. Ward, M.O. Wascko, G. Weiglein, D.H. Weinberg, E.J. Weinberg, M. White, L.R. Wiencke, S. Willocq, C.G. Wohl, J. Womersley, C.L. Woody, R.L. Workman, W.-M. Yao, G.P. Zeller, O.V. Zenin, R.-Y. Zhu, S.-L. Zhu, F. Zimmermann, P.A. Zyla, J. Anderson, L. Fuller, V.S. Lugovsky, and P. Schaffner. Review of particle physics. Phys. Rev. D **98**, 030001 (2018). https://doi.org/10.1103/PhysRevD.98.030001

53. C.Y. Seng, M. Gorchtein, M.J. Ramsey-Musolf, Dispersive evaluation of the inner radiative correction in neutron and nuclear β decay. Phys. Rev. D **100**(1), 013001 (2019). https://doi.org/10.1103/PhysRevD.100.013001
54. C.-Y. Seng, M. Gorchtein, H.H. Patel, M.J. Ramsey-Musolf, Reduced hadronic uncertainty in the determination of v_{ud}. Phys. Rev. Lett. **121**(24), 241804 (2018). https://doi.org/10.1103/PhysRevLett.121.241804
55. C.-Y. Seng, Radiative corrections to semileptonic beta decays: progress and challenges. Particles **4**(4), 397–467 (2021). https://doi.org/10.3390/particles4040034
56. J.C. Bergstrom, I.P. Auer, R.S. Hicks, Electroexcitation of the 0+ (3.562-MeV) Level of Li-6 and Its Application to the Reaction Li-6 (gamma, pi+) He-6. Nucl. Phys. A **251**, 401–417 (1975). https://doi.org/10.1016/0375-9474(75)90537-0
57. L. Hayen, A.R. Young, Consistent description of angular correlations in β decay for Beyond Standard Model physics searches. 9 (2020)
58. W.E. Kleppinger, F.P. Calaprice, and Barry R. Holstein. Analysis of the recoil ion spectra for the β-decays of 6He, 19Ne and 35Ar. *Nuclear Physics A*, 293(1-2):46–60, dec 1977. ISSN 03759474. https://doi.org/10.1016/0375-9474(77)90475-4. URL https://linkinghub.elsevier.com/retrieve/pii/0375947477904754
59. B.R. Holstein, *Weak Interactions in Nuclei*. Princeton University Press, 1989. ISBN 9781400887040
60. R.J. Blin-Stoyle, An introduction to exchange effects in nuclei. Prog. Part. Nucl. Phys. **1**(2), 5–40 (1978)
61. W. Buchmuller, D. Wyler, Effective lagrangian analysis of new interactions and flavor conservation. Nucl. Phys. B **268**, 621–653 (1986). https://doi.org/10.1016/0550-3213(86)90262-2
62. B. Grzadkowski, M. Iskrzynski, M. Misiak, J. Rosiek, Dimension-six terms in the standard model lagrangian. JHEP **10**, 085 (2010). https://doi.org/10.1007/JHEP10(2010)085
63. A. Glick-Magid, D. Gazit, Multipole decomposition of tensor interactions of fermionic probes with composite particles and bsm signatures in nuclear reactions. Phys. Rev. D **107**(7), 075031 (2023). https://doi.org/10.1103/PhysRevD.107.075031
64. Y. Aoki et al., Flag review 2021. Eur. Phys. J. C **82**(10), 869 (2022). https://doi.org/10.1140/epjc/s10052-022-10536-1
65. V. Cirigliano, S. Gardner, B. Holstein, Beta decays and non-standard interactions in the lhc era. Prog. Part. Nucl. Phys. **71**, 93–118 (2013). https://doi.org/10.1016/j.ppnp.2013.03.005
66. A. Falkowski, M. González-Alonso, O. Naviliat-Cuncic, Comprehensive analysis of beta decays within and beyond the standard model. JHEP **04**, 126 (2021). https://doi.org/10.1007/JHEP04(2021)126
67. S. Alioli, W. Dekens, M. Girard, E. Mereghetti, Nlo qcd corrections to sm-eft dilepton and electroweak higgs boson production, matched to parton shower in powheg. JHEP **08**, 205 (2018). https://doi.org/10.1007/JHEP08(2018)205
68. R. Boughezal, E. Mereghetti, F. Petriello, Dilepton production in the smeft at o(1/λ4). Phys. Rev. D **104**(9), 095022 (2021). https://doi.org/10.1103/PhysRevD.104.095022
69. F.P. Calaprice, D.J. Millener, Heavy neutrinos and the beta spectra of s-35, f-18, and ne-19. Phys. Rev. C **27**, 1175–1181 (1983). https://doi.org/10.1103/PhysRevC.27.1175

70. V.I. Isakov, M.I. Strikman, Restrictions on the mixing of electron and heavy neutrinos from the super allowed O+ –> O+ fermi beta decays. Phys. Lett. B **181**, 195–198 (1986). https://doi.org/10.1016/0370-2693(86)90030-4
71. J. Deutsch, M. Lebrun, R. Prieels, Searches for admixture of massive neutrinos into the electron flavor. Nucl. Phys. A **518**, 149–155 (1990). https://doi.org/10.1016/0375-9474(90)90541-S
72. A.V. Derbin, I.S. Drachnev, I.S. Lomskaya, V.N. Muratova, N.V. Pilipenko, D.A. Semenov, L.M. Tukkhonen, E.V. Unzhakov, A.K. Khusainov, ^{144}ce$-^{144}$pr. JETP Lett. **108**(8), 499–503 (2018). https://doi.org/10.1134/S0021364018200067
73. P.D. Bolton, F.F. Deppisch, P.S. Bhupal Dev, Neutrinoless double beta decay versus other probes of heavy sterile neutrinos. JHEP **03**, 170 (2020). https://doi.org/10.1007/JHEP03(2020)170
74. B. Dasgupta, J. Kopp, Sterile Neutrinos. Phys. Rept. **928**, 1–63 (2021). https://doi.org/10.1016/j.physrep.2021.06.002

Chapter 5
Partial Muon Capture Rates for $A = 3$ and $A = 6$ Nuclei

Abstract Neutrinoless double beta ($0\nu\beta\beta$) decay is a process where two neutrons in a nucleus are converted into two protons with the emission of two electrons and no neutrinos, thereby violating lepton number conservation by two units. The next generation of $0\nu\beta\beta$ decay experiments holds the potential to uncover the origin of the neutrino masses and the observed matter-antimatter asymmetry in the universe [1–4]. The observation of this decay would imply that neutrinos are Majorana particles and serve as a probe for several lepton-number-violating (LNV) mechanisms [1, 2], including the standard light-Majorana neutrino exchange. Regardless of the LNV decay mechanism, the interpretation of new physics hinges on an accurate understanding of the underlying nuclear dynamics. Rates of these decays depend not only on unknown parameters associated with BSM physics, but also on nuclear matrix elements that must be obtained from many-body approaches. These matrix elements can vary by a factor of two to three depending on the approach used [3, 4]. Therefore, improving the theoretical description of this process is crucial for the future of fundamental symmetries research. If one assumes that $0\nu\beta\beta$ decay results come from the exchange of a light Majorana neutrino between two nucleons, then the momentum carried by the neutrino is on the order of 100 MeV/c [3, 5]. Muon captures on nuclei involve momentum transfers on the order of the muon mass, thus placing it in a similar kinematic regime. Therefore, muon capture is a great testing ground to validate models of nuclear electroweak physics that will be relevant to studying $0\nu\beta\beta$-decay. Muon capture reactions have been treated extensively from both the theoretical and experimental points of view [6–9] and rates have been obtained in light systems with several methods [10–21]. In the following section, I will briefly review the theory of muon capture in light nuclei and discuss recent QMC calculations of partial muon capture rates on the ^{3}He and ^{6}Li ground states.

5.1 Theoretical Background

Muon capture on the atomic nucleus is a process in which a muon of mass m_μ with momentum $\mathbf{k}_\mu$ and energy E_μ and a proton in the nucleus of mass m_p are converted into a neutron of mass m_n and a neutrino of mass $m_\nu \simeq 0$ with momentum $\mathbf{k}_\nu$ and

G. B. King, *Electroweak Structure and Reactions in Light Nuclei with Quantum Monte Carlo Methods*, Springer Theses, https://doi.org/10.1007/978-3-032-03191-4_5

energy $E_\nu \simeq |\mathbf{k}_\nu|$. Assuming a muon at rest in a Hydrogen-like $1s$ oribtal and using the standard multipole analysis (see Sect. B.3), the partial rate to a single final state rate becomes [22]

$$\Gamma = \frac{G_F^2 V_{ud}^2}{2\pi} \frac{4\pi}{2J_i + 1} \mathcal{R} |\psi_{1s}(0)|^2 \frac{E_\nu^{*2}}{\text{recoil}} \times \left[|C_1(E_\nu^*; A) - L_1(E_\nu^*; A)|^2 + |E_1(E_\nu^*; A) - M_1(E_\nu^*; V)|^2 \right], \tag{5.1}$$

where the outgoing neutrino energy is given by [11],

$$E_\nu^* = \frac{\left(m_i + E_i + m_\mu\right)^2 - \left(m_f + E_f\right)^2}{2\left(m_i + E_i + m_\mu\right)}, \tag{5.2}$$

and the recoil factor is given by,

$$\frac{1}{\text{recoil}} = \left(1 - \frac{E_\nu^*}{m_i + E_i + m_\mu}\right). \tag{5.3}$$

The intial and final state energies of the nuclear systems are denoted as E_i and E_f, respectively, while

$$\begin{aligned} m_i &= Z\, m_p + N\, m_n + (Z-1)\, m_e\,, \\ m_f &= (Z-1)(m_p + m_e) + (N+1)\, m_n\,, \end{aligned} \tag{5.4}$$

for an initial nucleus with charge number Z and neutron number N. The expression also includes a reduction factor,

$$\mathcal{R} = \frac{|\psi_{1s}^{\text{av}}|^2}{|\psi_{1s}(0)|^2}, \tag{5.5}$$

which indicates how much the square of the Hydrogen-like $1s$ muon wave function averaged over the charge distribution of the nucleus ψ_{1s}^{av} differs from values of the square at the origin $\psi_{1s}(0)$. In some cases, it is more convenient to cast this rate in terms of matrix elements. Making the choice $\hat{\mathbf{k}}_\nu = -\hat{\mathbf{q}} = -\hat{\mathbf{z}}$, the rate is given as,

$$\begin{aligned} \Gamma = \frac{G_F^2 V_{ud}^2}{2\pi} \mathcal{R} \frac{|\psi_{1s}(0)|^2}{(2J_i + 1)} \frac{E_\nu^{*2}}{\text{recoil}} \times \\ \sum_{M_f, M_i} \Big[|\langle J_f, M_f | \rho(E_\nu^*, \hat{\mathbf{z}}) | J_i, M_i \rangle|^2 + |\langle J_f, M_f | j_z(E_\nu^* \hat{\mathbf{z}}) | J_i, M_i \rangle|^2 \\ +2\text{Re}\left[\langle J_f, M_f | \rho(E_\nu^* \hat{\mathbf{z}}) | J_i, M_i \rangle \langle J_f, M_f | j_z(E_\nu^* \hat{\mathbf{z}}) | J_i, M_i \rangle^* \right] \\ +|\langle J_f, M_f | j_x(E_\nu^* \hat{\mathbf{z}}) | J_i, M_i \rangle|^2 + |\langle J_f, M_f | j_y(E_\nu^* \hat{\mathbf{z}}) | J_i, M_i \rangle|^2 \\ -2\,\text{Im}\left[\langle J_f, M_f | j_x(E_\nu^* \hat{\mathbf{z}}) | J_i, M_i \rangle \langle J_f, M_f | j_y(E_\nu^* \hat{\mathbf{z}}) | J_i, M_i \rangle^* \right] \Big]. \end{aligned} \tag{5.6}$$

5.2 QMC Partial Muon Capture Rates

King et al. [23] presented results of the VMC and GFMC calculations of the partial muon capture rate on the ^{3}He and ^{6}Li ground states using the NV2+3 nuclear Hamiltonians In Table 5.1, I present the VMC results for the partial capture rates using all eight NV2+3 model classes. I performed the calculations using leading order one-body operators only [Γ(1b)] and with one- plus two-body electroweak currents [Γ(2b)]. The partial capture rate on ^{3}He has been precisely measured [24] and the one-body contribution alone cannot reproduce this measurement. With the two-body electroweak currents included, the VMC rates increase by about 9% to 16%. In the VMC calculation, the agreement with the datum ranges from about 0.1% to 4.6%. The $3N$ interaction fit has the most significant impact on the rate, leading to differences on average of 54 s^{-1} between models with the same NV2 but different NV3 potentials. This, once again, is attributable to the LEC c_D entering the $3N$ interaction which governs the strength of the axial CT current at N3LO in the chiral expansion [25]. The enhanced CT contribution in the unstarred models leads to large variations in the partial capture rate compared to the starred counterparts. The cutoff and energy range of the fit lead to changes of 16 s^{-1} and 22 s^{-1} on average, respectively, which is consistent with the findings of Refs. [12, 15].

Table 5.1 VMC calculations of partial muon capture rates in ^{3}He and ^{6}Li obtained with chiral one-body only [Γ(1b)], and one- and two-body [Γ(2b)] axial and vector currents corresponding to the eight NV2+3 models. The third column gives the factor $\mathcal{R}$ used to account for the finite nuclear charge distribution. The experimental result is given in the last column. All uncertainties on the theoretical predictions are Monte Carlo errors

Capture	Model	$\mathcal{R}$	Γ(1b) (s^{-1})	Γ(2b) (s^{-1})	Expt. (s^{-1})
$^3\text{He}(\frac{1}{2}^+;\frac{1}{2}) \rightarrow$ $^3\text{H}(\frac{1}{2}^+;\frac{1}{2})$	Ia (Ib)	0.995 (0.995)	1350.3 ± 0.8 (1363.4 ± 0.2)	1564.4 ± 0.9 (1545.7 ± 0.3)	1496.0 ± 4.0 [24]
	Ia* (Ib*)	0.995 (0.995)	1357.4 ± 0.2 (1358.5 ± 0.2)	1473.9 ± 0.3 (1483.6 ± 0.3)	
	IIa (IIb)	0.995 (0.995)	1369.7 ± 0.2 (1372.8 ± 0.2)	1533.8 ± 0.3 (1512.4 ± 0.3)	
	IIa* (IIb*)	0.995 (0.995)	1364.5 ± 0.2 (1372.5 ± 0.2)	1484.4 ± 0.3 (1497.0 ± 0.3)	
$^6\text{Li}(1^+;0) \rightarrow$ $^6\text{He}(0^+;1)$	Ia (Ib)	0.990 (0.990)	1196 ± 2 (1243 ± 2)	1282 ± 2 (1331 ± 2)	1600^{+330}_{-129} [26]
	Ia* (Ib*)	0.990 (0.990)	1154 ± 3 (1188 ± 2)	1177 ± 3 (1233 ± 2)	
	IIa (IIb)	0.990 (0.990)	1227 ± 2 (1142 ± 2)	1294 ± 2 (1185 ± 2)	
	IIa* (IIb*)	0.990 (0.990)	1215 ± 2 (1151 ± 2)	1257 ± 2 (1185 ± 2)	

In the ^{6}Li partial capture rate, the inclusion of two-body electroweak currents also increases the rate. Further, these currents again generate a greater enhancement in the non-starred models relative to their starred counterparts. Even with this enhancement, ranging approximately from 3% to 7%, the rates predicted at the VMC level for the NV2+3 models are about 11–21% slower than the available experimental datum [26]. In this case, the $3N$ fit is no longer the dominant contribution to the uncertainty. Rather, all sources contribute a similar model uncertainty. On average, the cutoff and energy range of the fit both change the rate by 72 s^{-1}, while the $3N$ interaction changes the rate by 60 s^{-1}.

In addition to the individual model predictions, I also computed an average VMC rate for both captures under study. To estimate the uncertainty on the rate, a similar procedure was used to that outlined in Sect. 4.3.2. By using the average changes due NV2 interaction cutoffs, the energy range used to fit the interaction, and the $3N$ fit, I assigned a total error estimate on the rate. An additional source of uncertainty was considered by varying the nucleonic axial radius parameter by $\pm 10\%$. This changes the rate by $\pm 0.6\%$. The four sources of uncertainty are combined in quadrature to determine the overall uncertainty on the VMC averages, obtaining,

$$\begin{aligned}\Gamma(A{=}3;\,\mathrm{VMC}) &= 1512 \pm 32\ \mathrm{s}^{-1},\\ \Gamma(A{=}6;\,\mathrm{VMC}) &= 1243 \pm 59\ \mathrm{s}^{-1}.\end{aligned}$$

To remove spurious contamination from the wave functions for models Ia and Ia*, I performed a GFMC calculation and the corresponding results are reported in Table 5.2. These two models provided the fastest and slowest VMC partial capture rates for $A = 3$ and thus make a rough upper and lower bound on the GFMC rates. Figure 5.1 displays the average VMC results, along with the VMC and GFMC results for models Ia and Ia*, compared with experimental data and past theoretical calculations. An average GFMC value is also cited, with its error taken to be half the difference between the Ia and Ia* rates.

At the VMC level, model Ia overpredicted the $A = 3$ muon capture rate by 4.6%. After propagation, that rate was decreased and reaches a 1.5% agreement with the datum. By contrast, model Ia*, which had 1.4% agreement with the experimental

Table 5.2 VMC and GFMC calculations of partial muon capture rates in ^{3}He and ^{6}Li obtained with chiral one- and two-body axial and vector currents with the NV2+3 models. The experimental result is given in the last column. All uncertainties on the theoretical predictions are Monte Carlo errors

Capture	Model	Γ(VMC) (s^{-1})	Γ(GFMC) (s^{-1})	Expt
$^3\mathrm{He}(\frac{1}{2}^+;\frac{1}{2}) \to {}^3\mathrm{H}(\frac{1}{2}^+;\frac{1}{2})$	Ia	1564.4 ± 0.9	1519 ± 3	1496.0 ± 4.0 [24]
	Ia*	1473.9 ± 0.3	1433 ± 2	
$^6\mathrm{Li}(1^+;0) \to {}^6\mathrm{He}(0^+;1)$	Ia	1282 ± 2	1277 ± 10	$1600\,^{+330}_{-129}$ [26]
	Ia*	1177 ± 2	926 ± 8	

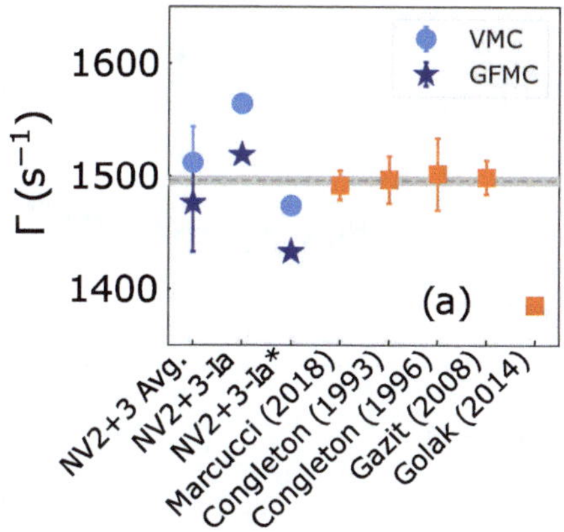

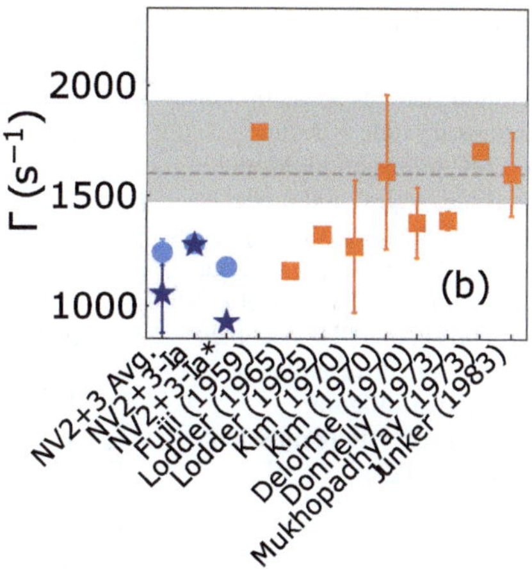

Fig. 5.1 The partial muon capture rate in **a** ^{3}He and **b** ^{6}Li from the NV2+3-Ia and NV2+3-Ia* models in VMC (light blue circle) and GFMC (dark blue star) calculations compared with other work (orange squares) [12–15, 27–33]. The experimental values (dashed gray line) and their error (shaded region) [24, 26] are included for comparison with the theory predictions. Figure reproduced with permission from Ref. [23]

datum at the VMC level, now underpredicts the rate by 4.2%. The average GFMC rate of 1519±45 agrees with data and past χEFT calculations containing one- and two-body currents within model uncertainty. In Fig. 5.1 panel (a), the past χEFT calculations are represented by Refs. [12] and [15] for calculations with many-body current corrections, and by Ref. [16] for a purely one-body calculation. Even when scaled by the more recent value of G_V, the rate of Ref. [12] falls within the GFMC band. Though there is clearly agreement, because these past calculations use a different set of currents and underlying nuclear interactions than the present work, it is difficult to directly compare them to the GFMC results. Thus, in the future, benchmark calculations with other many-body methods based on the same dynamical inputs would be useful to further validate the microscopic approach adopted in Ref. [23]

While both $A = 3$ GFMC rates exhibit few-percent decreases from the VMC result, the $A = 6$ rates display a dramatically different behavior for models Ia and Ia*. The matrix elements for the model Ia calculation were fairly stable when propagated from VMC to GFMC, resulting in a modest sub-percent change of the overall rate; however, for model Ia*, the dominant matrix elements changed at the few percent level, but since the rate is proportional to the square of the matrix element, this leads to a change of roughly 20% in the rate once the matrix element is converged.

The origin of the large change in the model Ia* matrix element is based on the same argument as for why the model IIa* matrix elements relevant to ^{6}He β-decay were affected by a large systematic uncertainty in Sect. 4.3.2. Once again, it is possible to compare the system size as a function of τ during the GFMC propagation of the $A = 6$ nuclei for the two models. The system size for ^{6}Li$(1^+; 0)$ grows at the

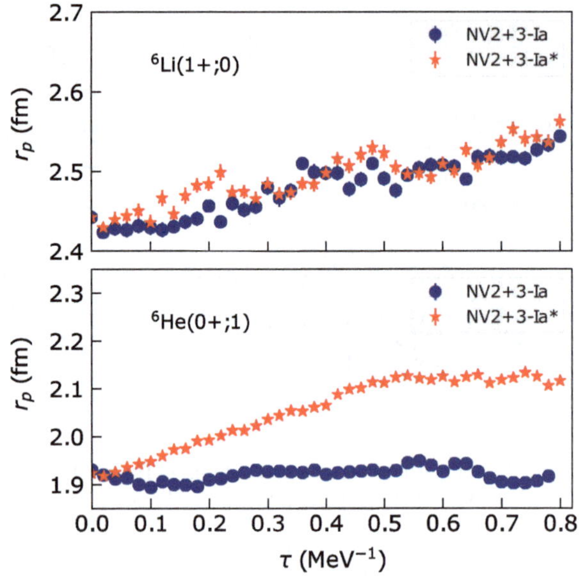

Fig. 5.2 The point proton rms radii for the ^{6}Li and ^{6}He ground states as a function of imaginary time τ during GFMC propagation. Model Ia (blue circles) and model Ia* (orange stars) display similar behaviors for ^{6}Li, but differ for ^{6}He. The model Ia* ground state of ^{6}He becomes much more diffuse than its counterpart. Figure reproduced with permission from Ref. [23]

same rate in τ for both models; however, the ^{6}He(0^+; 1) ground state size is stable for model Ia, while it increases monotonically in τ before beginning to converge for model Ia* (see Fig. 5.2). Because of the $e^{-i\mathbf{q}\cdot\mathbf{r}_i}$ dependence in the dominant one-body terms of the current operator, the matrix elements at a finite value of q experience a more significant drop for model Ia* due to the diffuseness of ^{6}He(0^+; 1) with that interaction. Performing the same analysis for the $A = 3$ system shows that the system size is consistent between both models as a function of τ, explaining why they change by roughly the same amount. Given the sensitivity to the size of the system, it is critical that microscopic calculations are able to reproduce this quantity.

The difference with experiment in $A = 6$ is significant for both models Ia and Ia*, especially when compared with the few percent agreement obtained in GFMC calculations of the GT matrix element for the ^{6}He $\rightarrow$ ^{6}Li beta decay [25] and for the ^{6}Li magnetic moment (see Sect. 6.2). As detailed in Ref. [33], calculations of this rate [27–33] have ranged from 1160 s^{-1} to 1790 s^{-1}. The calculation of Ref. [33] matched the experimental datum by modelling ^{6}Li as a ^{3}He + t cluster and using the Fujii-Primakoff [27] effective Hamiltonian for muon capture. Sub-percent agreement was also obtained by Ref. [30], which treated the ^{6}Li and ^{6}He nuclei as elementary particles with magnetic and axial form factors extracted from experimental data. The two calculations presented by the authors of that work adopted different formulations of the partially conserved axial current (PCAC) relation to obtain the pseudoscalar form factor, with the faster rate using the Gell-Mann-Lévy version [34] and the slower rate using the Nambu one [35]. The Nambu definition is consistent with the induced pseudoscalar term in the weak axial current from χEFT.

It is difficult to compare the QMC results with those of previous theoretical treatments of the ^{6}Li partial capture rate, as most do not treat the nucleus in a microscopic picture. For example, in the work of Ref. [31], the ^{6}Li and ^{6}He bound states are described by shell-model wave functions with valence configurations restricted to the $1p$-shell; moreover, the nuclear electroweak current neglects meson-exchange contributions. The QMC result at leading order (obtained with one-body currents) is quenched relative to the shell model one, as one would have expected. Given this fact and the agreement between with the analysis of Ref. [30] based on experimental data using the same pseudoscalar form factor, it is reasonable to be confident in this result. Recent NCSM results including two-body currents in an approximate way agree with the QMC result and further motivate other exact treatments of the capture rate [36]. More modern calculations with other many-body methods and a novel measurement of the rate would be valuable in establishing the validity of the nuclear inputs and many-body approach. Further, a more precise measurement against which microscopic calculations could benchmark electroweak current models would be useful to the nuclear theory community.

References

1. V. Cirigliano et al., Towards precise and accurate calculations of neutrinoless double-beta decay. J. Phys. G **49**(12), 120502 (2022). https://doi.org/10.1088/1361-6471/aca03e
2. V. Cirigliano et al., Neutrinoless double-beta decay: A roadmap for matching theory to experiment. 3 (2022b)
3. J. Engel, J. Menéndez, Status and future of nuclear matrix elements for neutrinoless double-beta decay: A review. Rept. Prog. Phys. **80**(4), 046301 (2017). https://doi.org/10.1088/1361-6633/aa5bc5
4. M. Agostini, G. Benato, J.A. Detwiler, J. Menéndez, F. Vissani, Toward the discovery of matter creation with neutrinoless $\beta\beta$ decay. Rev. Mod. Phys. **95**(2), 025002 (2023). https://doi.org/10.1103/RevModPhys.95.025002
5. S. Pastore, J. Carlson, V. Cirigliano, W. Dekens, E. Mereghetti, R.B. Wiringa, Neutrinoless double-β decay matrix elements in light nuclei. Phys. Rev. C **97**(1), 014606 (2018). https://doi.org/10.1103/PhysRevC.97.014606
6. N.C. Mukhopadhyay, Nuclear muon capture. Phys. Rep. **30**(1), 1 – 144 (1977). ISSN 0370-1573. https://doi.org/10.1016/0370-1573(77)90073-4. http://www.sciencedirect.com/science/article/pii/0370157377900734
7. D.F. Measday, The nuclear physics of muon capture. Phys. Rep. **354**(4), 243 – 409 (2001). ISSN 0370-1573. https://doi.org/10.1016/S0370-1573(01)00012-6. http://www.sciencedirect.com/science/article/pii/S0370157301000126
8. P. Kammel, K. Kubodera, Precision muon capture. Annu. Rev. Nucl. Part. Sci. **60**(1), 327–353 (2010). https://doi.org/10.1146/annurev-nucl-100809-131946
9. T.P. Gorringe, D.W. Hertzog, Precision muon physics. Progr. Particle Nuclear Phys. **84**, 73–123 (2015). ISSN 01466410. https://doi.org/10.1016/j.ppnp.2015.06.001. URL https://linkinghub.elsevier.com/retrieve/pii/S0146641015000435
10. L.E. Marcucci, R. Schiavilla, S. Rosati, A. Kievsky, M. Viviani, Theoretical study of the He-3(mu-, muon-neutrino) H-3 capture. Phys. Rev. C **66**, 054003 (2002). https://doi.org/10.1103/PhysRevC.66.054003

11. L.E. Marcucci, M. Piarulli, M. Viviani, L. Girlanda, A. Kievsky, S. Rosati, R. Schiavilla, Muon capture on deuteron and ^{3}He. Phys. Rev. C **83**, 014002 (2011). https://doi.org/10.1103/PhysRevC.83.014002
12. L.E. Marcucci, A. Kievsky, S. Rosati, R. Schiavilla, M. Viviani, Chiral effective field theory predictions for muon capture on deuteron and ^3He. Phys. Rev. Lett. **108**, 052502 (2012). 10.1103/PhysRevLett.121.049901, https://doi.org/10.1103/PhysRevLett.108.052502. [Erratum: Phys. Rev. Lett.121, no.4,049901(2018)]
13. J.G. Congleton, H.W. Fearing, Determination of the nucleon pseudoscalar coupling using muon capture by 3he. Nuclear Phys. A **552**(4), 534 – 548 (1993). ISSN 0375-9474. https://doi.org/10.1016/0375-9474(93)90284-5. http://www.sciencedirect.com/science/article/pii/0375947493902845
14. J.G. Congleton, E. Truhlík, Nuclear muon capture by ^{3}He: Meson exchange currents for the triton channel. Phys. Rev. C **53**, 956–976 (1996). https://doi.org/10.1103/PhysRevC.53.956
15. D. Gazit, The weak structure of the nucleon from muon capture on 3he. Nuclear Phys. A **827**(1), 408c – 410c (2009). ISSN 0375-9474https://doi.org/10.1016/j.nuclphysa.2009.05.087. http://www.sciencedirect.com/science/article/pii/S0375947409003510. PANIC08
16. J. Golak et al., Low-energy neutron-deuteron reactions with N3LO chiral forces. Eur. Phys. J. A **50**, 177 (2014). https://doi.org/10.1140/epja/i2014-14177-7
17. J. Adam, M. Tater, E. Truhlík, E. Epelbaum, R. Machleidt, P. Ricci, Calculation of doublet capture rate for muon capture in deuterium within chiral effective field theory. Physics Letters, Section B: Nuclear, Elementary Particle and High-Energy Physics, **709**(1-2), 93–100 (2012). ISSN 03702693. https://doi.org/10.1016/j.physletb.2012.01.065
18. E. Kolbe, K. Langanke, P. Vogel, Muon capture, continuum random phase approximation, and in-medium renormalization of the axial-vector coupling constant. Phys. Rev. C **50**, 2576–2581 (1994). https://doi.org/10.1103/PhysRevC.50.2576
19. T. Marketin, N. Paar, T. Nikšć, D. Vretenar, Relativistic quasiparticle random-phase approximation calculation of total muon capture rates. Phys. Rev. C **79**, 054323 (2009). https://doi.org/10.1103/PhysRevC.79.054323
20. A.R. Samana, F. Krmpoti ć, N. Paar, C.A. Bertulani, Neutrino and antineutrino charge-exchange reactions on ^{12}C. Phys. Rev. C **83**, 024303 (2011). https://doi.org/10.1103/PhysRevC.83.024303
21. P.G. Giannaka, T.S. Kosmas, Detailed description of exclusive muon capture rates using realistic two-body forces. Phys. Rev. C **92**, 014606 (2015). https://doi.org/10.1103/PhysRevC.92.014606
22. J.D. Walecka, *Theoretical nuclear and subnuclear physics* (Oxford University Press, New York, 1995)
23. G.B. King, S. Pastore, M. Piarulli, R. Schiavilla, Partial muon capture rates in A=3 and A=6 nuclei with chiral effective field theory. Phys. Rev. C **105**(4), L042501 (2022)
24. P. Ackerbauer, D.V Balin, V.M Baturin, G.A Beer, W.H Breunlich, T Case, K Crowe, H Daniel, J Deutsch, J Govaerts, Yu.S Grigoriev, F.J Hartmann, P Kammel, R King, B Lauss, E.M Maev, V.E Markushin, J Marton, M Mühlbauer, C Petitjean, Th Petitjean, G.E Petrov, R Prieels, W Prymas, W Schott, G.G Semenchuk, Yu.V Smirenin, A.A Vorobyov, N.I. Voropaev, and P Wojciechowski. A precision measurement of nuclear muon capture on 3he. Phys. Lett. B **417**(3), 224–232 (1998). ISSN 0370-2693. https://doi.org/10.1016/S0370-2693(97)01382-8. http://www.sciencedirect.com/science/article/pii/S0370269397013828
25. G.B. King, L. Andreoli, S. Pastore, M. Piarulli, R. Schiavilla, R.B. Wiringa, J. Carlson, S. Gandolfi, Chiral effective field theory calculations of weak transitions in light nuclei. Phys. Rev. C **102**, 025501 (2020). https://doi.org/10.1103/PhysRevC.102.025501
26. J.P. Deutsch, L. Grenacs, P. Igo-Kemenes, P. Lipnik, P.C. Macq, An experimental test of the analogy between muon and radiative pion capture. Phys. Lett. B **26**(5), 315–316 (1968). ISSN 0370-2693. https://doi.org/10.1016/0370-2693(68)90571-6. http://www.sciencedirect.com/science/article/pii/0370269368905716
27. A. Fujii, H. Primakoff, Muon capture in certain light nuclei. Il Nuovo Cimento (1955–1965) **12**(4), 327–355 (1959)

28. A. Lodder, C.C. Jonker, Muon capture in li6 as a tool for testing nuclear models. Phys. Lett. **15**(3), 245–248 (1965)
29. J. Delorme, On the elementary-particle treatment of muon capture and radiative pion capture in 6li. Nucl. Phys. B **19**(2), 573–588 (1970)
30. C.W. Kim, S.L. Mintz, Muon-capture rate in 6li and the pcac hypothesis. Phys. Lett. B **31**(8), 503–505 (1970)
31. T.W. Donnelly, J.D. Walecka, Electromagnetic and weak interactions with nuclei: One-body densities in 6li. Phys. Lett. B **44**(4), 330–334 (1973)
32. N.C. Mukhopadhyay, Giant magnetic dipole analog excitations by the allowed muon capture in 1p-shell nuclei. Phys. Lett. B **45**(4), 309–312 (1973)
33. K. Junker, A study of the partial muon capture rates μ-+ 6li 6he (gs)+ $\nu\mu$ and μ-+ 6li t+ t+ $\nu\mu$. Nucl. Phys. A **407**(3), 460–494 (1983)
34. M. Gell-Mann, M. Levy, The axial vector current in beta decay. Nuovo Cim. **16**, 705 (1960). https://doi.org/10.1007/BF02859738
35. Y. Nambu, Axial vector current conservation in weak interactions. Phys. Rev. Lett. **4**, 380–382 (1960). https://doi.org/10.1103/PhysRevLett.4.380
36. L. Jokiniemi, P. Navratil, J. Kotila, K. Kravvaris, Muon Capture on ^{6}Li, ^{12}C, and ^{16}O from *Ab Initio* Nuclear Theory. 3 (2024)

Chapter 6
Magnetic Moments and Form Factors

Abstract The magnetic moment of a nucleus describes the strength and orientation of the magnetic field produced by the electric current density inside of the nucleus. The proton carries charge and both species of nucleon have non-zero gyromagnetic ratios, thus in the single-particle picture, the nuclear magnetic moment arises from the convection current associated with the motion of the protons and from the spin magnetization of the nucleons. Often times, however, the simple single-particle model fails to explain the experimental data. This is not necessarily surprising, however, since the potential between the protons and neutrons has an exchange component and therefore, can move charge throughout the nucleus. In Ref. [1], the effect of two-body currents was analyzed in QMC studies of the magnetic moments of light nuclei using both phenomenological and hybrid approaches. These studies indicated the need for many-nucleon currents in order to reach agreement with experimental data. In this section, I present a study of magnetic moments using a consistent the χEFT framework provided by the NV2+3 models.

6.1 Multipole Analysis

The magnetic form factor of a nucleus is defined in terms of reduced multipoles as [2],

$$F_M^2(q) = \frac{1}{2J+1} \sum_{L=1}^{\infty} |\langle J||M_L(q)||J\rangle|^2 \,. \tag{6.1}$$

As was done for the ^{6}He β-decay spectrum, the low-q behavior of the multipoles may be leveraged in order to obtain a simple expression of the $M_1(q)$. Considering also that higher order multipoles will be negligible at small q compared with the $L = 1$ component, the following is true in that limit [3],

$$\langle J||M_1(q)||J\rangle \simeq \frac{i}{\sqrt{2\pi}\langle JJ, J-J|10\rangle} \frac{q}{2m_N}\mu \,. \tag{6.2}$$

G. B. King, *Electroweak Structure and Reactions in Light Nuclei with Quantum Monte Carlo Methods*, Springer Theses, https://doi.org/10.1007/978-3-032-03191-4_6

Taking the spin-quantization axis as $\hat{\mathbf{z}}$, one may choose $\hat{\mathbf{q}} = \hat{\mathbf{x}}$ to obtain the simple expression (see Eq. B.9),

$$\mu = \lim_{q\to 0} -i\frac{2m_N}{q}\langle J\, J\big|\mathrm{j}_y(q\hat{\mathbf{x}})\big|J\, J\rangle\,. \tag{6.3}$$

To extract μ, one evaluates $\langle J\, J\big|\mathrm{j}_y(q\hat{\mathbf{x}})\big|J\, J\rangle$ for $q \in [0.00, 0.25]$ fm^{-1} and fits the results to a polynomial, similarly to what was done for the multipole analysis of the ^{6}He β-decay spectrum in Ref. [4]. The linear coefficient is then used to determine the value of μ.

6.2 VMC Results

Magnetic moment calculations were performed for several light nuclei using the VMC method with the NV2+3 interactions and currents in Ref. [9]. Here, I summarized the results of this study. Figure 6.1 displays the results for nuclei with $3 \leq A \leq 10$ using only the LO one-body operator (empty circles) and with currents up to N3LO in the chiral expansion (filled symbols) for the NV2+3-Ia* and NV2+3-IIb*. The subleading currents are essential to get good agreement with experiment

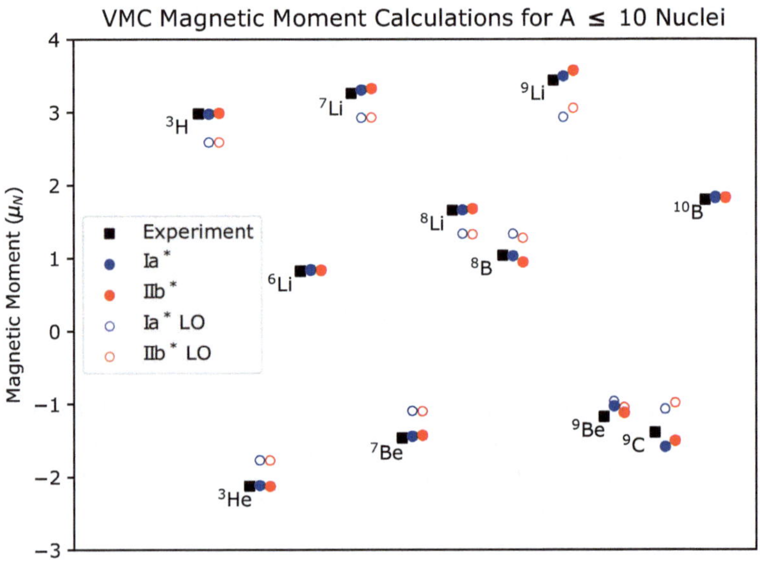

Fig. 6.1 Magnetic moments calculated with VMC using the N2+3-Ia* (blue circles) and N2+3-IIb* (red circles) compared with experiment (black dots). Empty symbols denote the calculation up to LO in the current, while filled symbols include all contributions through N3LO. The experimental data are from the evaluations of [5–8]. Figure reproduced with permission from Ref. [9]

and, in some cases, contribute ~30% of the total magnetic moment. The summed contributions to the matrix element displayed consistency across different model classes, as evidenced by the qualitative agreement in the figure between models, represented with blue and red symbols for the NV2+3-Ia⋆ and NV2+3-IIb⋆, respectively. The calculations also agree well with the experimental data shown by the black squares.

To further understand the calculations, an order-by-order analysis of the magnetic moment was performed. Table 6.1 reports these results for the VMC calculations performed with the various model classes. For nuclei with $3 \leq A \leq 7$, calculations are performed using four model classes of the Norfolk potential; namely, models Ia⋆, IIa⋆, Ib⋆, and IIb⋆. In this way, it is possible to assess how the cutoff and energy range used to fit the model impact magnetic moments. The results in Table 6.1 are broken down into different contributions listed in Table 2.1 and represented in Fig. 2.3; namely, I present the contribution from the one-body operator at LO in the chiral expansion, the OPE operator at NLO, the sum of the one-body relativistic correction and the OPE-range Δ-excitation diagram at N2LO, and in terms of the individual toplogies contributing at N3LO. Namely, the CT diagrams from minimal substitution into the NV2+3 (MIN), the non-minimal CT contribution (NM), the long-range OPE range contribution, and an intermediate range loop contribution involving the exchange of two pions. The coordinate space expressions of all operators are provided in Sect. A.2 of the Appendices.

As stated in Sect. 2.3, the trinucleon magnetic moment data are included in the fit of the unknown LECs in the subleading OPE and CT currents. Thus, the total trinucleon magnetic moments are reproduced by construction. At LO, the ^{3}H (^{3}He) results are stable across NV2+3 model classes. This is attributed to the [3] spatial symmetry of the nucleus (using the representation of the Young Scheme as described in Sect. 3.1), which requires that neutrons (protons) are spatially symmetric and thus that they occupy opposite spin states. Therefore, the magnetic moment of the entire nucleus at LO is essentially that of a free proton (neutron). The NLO and N2LO contributions, however, show a dependence on the choice of NV2+3 model. For both models, this contribution is much more sensitive to changing the cutoff than to changing the energy range used to fit the *NN* potential. At N3LO, the NM and OPE show a significant model dependence, likely due to the fact that the LECs appearing in these currents are fit to data. Similarly to Refs. [10, 11], it is observed that the N3LO(NM) contribution is enhanced relative to power counting expectations and that it gives a contribution that is similar in size to NLO. The other fit contribution, N3LO(OPE), generally gives a contribution with a size consistent with N2LO in the power counting except for model Ia⋆. A nearly exact cancellation between the two isovector (IV) contributions to the current leads to this accidental suppression of the matrix element. The N3LO(MIN) contribution appears to show a mild sensitivity to the energy range used to fit the potential while the N3LO(LOOP) has slight cutoff dependence. Overall, the model dependencies in the current cancel out in the summed matrix element, which varies by $\lesssim 0.5\%$ across all models for both magnetic moments. Again, the small variation of the summed matrix element is unsurprising given that the $A = 3$ magnetic moments are included in the fit to obtain the unknown LECs in the

Table 6.1 Magnetic moments in unit of μ_N computed with the NV2+3 interaction and consistent electromagnetic current broken down order-by-order in the chiral expansion. The column denoted by TOT is the sum of all contributions to the magnetic moment, while TOT-LO indicates the sub-leading current contribution to the result. Nuclei denoted with a dagger indicate states that were generated by swapping the numbers of protons and neutrons in the wave function optimized for the mirror system. Statistical uncertainties from the Monte Carlo integration, which are $\lesssim 2\%$ for all contributions listed, have been omitted. The experimental data are from the evaluations of [5–8]

	Model	LO	NLO	N2LO	N3LO(MIN)	N3LO(NM)	N3LO(OPE)	N3LO(LOOP)	TOT	TOT-LO
^{3}H($\frac{1}{2}^+$; $\frac{1}{2}$)	Ia*(IIa*)	2.588(2.588)	0.196(0.197)	0.036(0.035)	0.042(0.038)	0.089(0.125)	–0.005(–0.038)	0.027(0.027)	2.973(2.972)	0.385(0.384)
	Ib*(IIb*)	2.589(2.592)	0.224(0.226)	0.061(0.059)	0.042(0.033)	0.076(0.072)	–0.030(–0.014)	0.021(0.020)	2.972(2.986)	0.394(0.394)
	Exp								2.979	
^{3}He($\frac{1}{2}^+$; $\frac{1}{2}$)	Ia*(IIa*)	–1.766(–1.767)	–0.193(–0.194)	–0.042(–0.043)	0.030(0.029)	–0.117(–0.146)	–0.002(0.032)	–0.027(–0.026)	–2.116(–2.116)	–0.350(–0.358)
	Ib*(IIb*)	–1.770(–1.769)	–0.220(–0.222)	–0.066(–0.066)	0.033(0.026)	–0.124(–0.102)	0.045(0.026)	–0.026(–0.019)	–2.124(–2.127)	–0.358(–0.349)
	Exp								–2.127	
^{6}Li(1$^+$; 0)	Ia*(IIa*)	0.826(0.823)	0.000(0.000)	–0.010(–0.010)	0.046(0.043)	–0.019(–0.016)	–0.003(–0.002)	0.000(0.000)	0.840(0.838)	0.014(0.016)
	Ib*(IIb*)	0.820(0.825)	0.000(0.000)	–0.010(–0.010)	0.037(0.034)	–0.015(–0.019)	0.001(0.008)	0.000(0.000)	0.833(0.838)	0.015(0.13)
	Exp								0.822	
^{7}Li($\frac{3}{2}^-$; $\frac{1}{2}$)	Ia*(IIa*)	2.923(2.895)	0.184(0.187)	0.021(0.021)	0.046(0.043)	0.103(0.141)	–0.007(–0.038)	0.033(0.032)	3.301(3.280)	0.379(0.385)
	Ib*(IIb*)	2.905(2.928)	0.224(0.220)	0.052(0.043)	0.046(0.037)	0.092(0.083)	–0.032(–0.014)	0.029(0.026)	3.317(3.322)	0.413(0.394)
	Exp								3.256	
^{7}Be($\frac{3}{2}^-$; $\frac{1}{2}$)	Ia*(IIa*)	–1.094(–1.070)	–0.183(–0.187)	–0.0331(–0.0348)	0.030(0.032)	–0.132(–0.169)	0.002(0.032)	–0.033(–0.031)	–1.443(–1.465)	–0.349(–0.359)
	Ib*(IIb*)	–1.083(–1.098)	–0.222(–0.221)	–0.065(–0.060)	0.037(0.030)	–0.148(–0.119)	0.049(0.027)	–0.029(–0.025)	–1.461(–1.464)	–0.359(–0.367)
	Exp								–1.465	
^{8}Li(2$^+$; 1)	Ia*	1.332	0.183	0.016	0.064	0.080	–0.046	0.031	1.660	0.328

(continued)

Table 6.1 (continued)

	Model	LO	NLO	N2LO	N3LO(MIN)	N3LO(NM)	N3LO(OPE)	N3LO(LOOP)	TOT	TOT-LO
	IIb⋆	1.325	0.208	0.035	0.049	0.057	–0.018	0.022	1.678	0.353
	Exp								1.654	
$^8B(2^+;1)$	Ia⋆	1.330	–0.176	–0.047	0.054	–0.127	0.026	–0.032	1.028	–0.302
	IIb⋆	1.274	–0.206	–0.065	0.040	–0.108	0.031	–0.023	0.943	–0.331
	Exp								1.036	
$^9Li(\frac{3}{2}^-;\frac{3}{2})$	Ia⋆	2.933	0.271	0.031	0.094	0.113	0.026	0.027	3.494	
	IIb⋆	3.058	0.310	0.055	0.068	0.084	–0.005	0.006	3.574	0.517
	Exp								3.437	
$^9Be(\frac{3}{2}^-;\frac{1}{2})$	Ia⋆	–0.964	–0.041	0.011	0.010	–0.053	0.018	–0.010	–1.030	–0.066
	IIb⋆	–1.049	–0.049	0.004	0.011	–0.044	0.011	–0.007	–1.123	–0.074
	Exp								–1.178	
$^9B(\frac{3}{2}^-;\frac{1}{2})^\dagger$	Ia⋆	2.690	0.040	–0.024	0.024	0.039	–0.015	0.010	2.763	0.073
	IIb⋆	2.821	0.049	–0.017	0.015	0.029	–0.005	0.007	2.900	0.079
	Exp								–	
$^9C(\frac{3}{2}^-;\frac{3}{2})^\dagger$	Ia⋆	–1.070	–0.264	–0.047	0.002	–0.150	–0.033	–0.028	–1.590	–0.520
	IIb⋆	–0.983	–0.329	–0.079	0.003	–0.131	0.020	–0.007	–1.505	–0.523
	Exp								–1.391	
$^{10}B(3^+;0)$	Ia⋆	1.816	0.000	–0.019	0.069	–0.028	0.000	0.000	1.839	0.023
	IIb⋆	1.819	0.000	–0.019	0.048	–0.027	0.012	0.000	1.833	0.014
	Exp								1.801	

current. Two body currents contribute ~13% (16%) to the overall ^{3}H (^{3}He) magnetic moments.

Given the [43] spatial symmetry in $A = 7$ and the $\alpha + {}^3$H (^{3}He) cluster structure of ^{7}Li (^{7}Be), the order-by-order breakdown of the magnetic moments is approximately the same as for ^{3}H (^{3}He). Instead, ^{6}Li with its dominant [42] spatial symmetry, displays a strong $\alpha + d$ cluster structure. At leading order, the contribution to the magnetic moment is thus approximately the value of that for a free deuteron (μ_d = 0.857438 μ_N [12]) and displays little model dependence. Further, as ^{6}Li is a purely isoscalar (IS) ($T = 0$) nucleus, the matrix elements for IV components of the current operator vanish. Thus, some contributions from the current are suppressed beyond what one would expect from power counting considerations. For instance, the NLO contribution to the current is purely IV and thus does not contribute to the ^{6}Li magnetic moment. The N3LO(MIN) and N3LO(NM) contributions in this case are still enhanced, making few percent contributions to the overall magnetic moment. There is a ~15% model dependence on the N3LO(MIN) contribution that is driven by changing the cutoff, whereas the N3LO(NM) contribution is much more stable across models. The N3LO(OPE), contrary to its behavior for $A = 3$ and $A = 7$, contributes with a size that is consistent with its power counting. The N3LO(OPE) term also displays a large model dependence in ^{6}Li, though this term essentially makes no contribution to the overall matrix element. Again, the summed magnetic moment is stable across models with very little model dependence. Now, two-body currents only account for ~2% of the total magnetic moment.

Because of the cutoff dependence in the long-range NLO and N2LO terms and the general model dependence in the N3LO terms, I will focus on two representative model classes, Ia* and IIb*, for the heavier nuclei with $8 \leq A \leq 10$ to limit the number of computations. I will focus first on the purely IS ^{10}B ground state. As for ^{6}Li, the IV terms are suppressed relative to their expected power counting. The LO contribution in this case is once again rather stable, this time differing from the free d due to contributions from the orbital magnetic moments of the nucleons. The N3LO(MIN) and N3LO(OPE) contributions display model dependence and contributions that would be consistent with one order lower in the power counting. This model dependence again cancels out in the summed result and leads to a consistency for the two interactions. As for ^{6}Li, two-body currents only play a small role due to the suppression of IV terms and make up ~1% of the final value.

The addition of an unpaired p-shell neutron (proton) to ^{7}Li (^{7}Be) forms the predominantly [431] spatial symmetry ground state of ^{8}Li (^{8}B). At LO, the uncoupled valence nucleon quenches the magnitude of μ. Interestingly, the NLO and N2LO contributions are not greatly altered by this additional nucleon. In these nuclei, the extra binding from OPE that arises due to the unpaired nucleon is very little [13] and thus the one-pion range contributions at NLO and N2LO stay roughly the same in $A = 8$ as in $A = 7$. This carries over also into the long-range N3LO contributions, while the short-range N3LO contributions remain enhanced relative to their power counting and model dependent. For ^{8}B, two-body currents quench μ by ~30%. In ^{8}Li, they instead enhance μ by ~20%.

For $A = 9$, the structural differences of the $T = 3/2$ ground state of ^{9}Li and the $T = 1/2$ ground state of ^{9}Be lead to interesting physics in the two-body currents. As discussed in the previous QMC evaluation within the hybrid approach [1], the [432] spatial symmetry dominates the ^{9}Li ground state, giving it an $\alpha + {}^3\mathrm{H} + 2n$ cluster structure. On the other hand, ^{9}Be is dominated by a [441] spatial symmetry and clusters as $2\alpha + n$. A single valence neutron outside of the 2α cluster experiences no binding from OPE [13] and thus the NLO contribution is suppressed relative to its power counting. Before considering mixing from the tensor force, the ratio of p-shell IS pairs to IV pairs in ^{9}Be is enhanced by a factor of 2 relative to ^{9}Li. The IS pairs will not couple to the IV NLO two-body current, bringing down its contribution to μ in ^{9}Be relative to ^{9}Li. This suppression, due to purely nuclear structure effects, is responsible for the small two-body current contribution to μ of $\sim$6% in ^{9}Be. In ^{9}Li, there is binding from OPE felt both within the triton and between the triton and valence neutrons. Here, two-body currents play a much larger role and make a contribution of $\sim$15% to the value of μ. The wave functions of the mirrors to ^{9}Li and ^{9}Be– ^{9}C and ^{9}B, respectively– were generated by simply swapping the numbers of protons and neutrons without re-optimizing the variational parameters. Thus, for similar reasons, the two-body current contribution in ^{9}B is suppressed while in ^{9}C, it turns out to be significant and provides $\sim$33% of the total value of μ. As in the $A \leq 8$ nuclei, the N3LO contributions in $A = 9$ depend on the model. In the case of the $T = 3/2$ nuclei, the N3LO(NM) contribution is still roughly NLO while it is suppressed in $T = 1/2$ to a size that would roughly be consistent with N2LO in the power counting.

Analyzing the trends in the systematic order-by-order calculations of μ for several nuclei, the N3LO(NM) tends to be enhanced to roughly NLO for nuclei where IV currents do not vanish. In the IS case, the enhancement is only by approximately one order. Another standout case is $A = 9$, $T = 3/2$ where both the NLO(OPE) and N3LO(NM) terms are suppressed due to their $2\alpha + N$ cluster structures. It is also worth noting that the one-pion range terms at NLO and N2LO vary with the cutoff. An evaluation using two models is certainly not enough to claim a systematic cutoff dependence; however given the hints of cutoff sensitivity and the enhancement of CT terms fit to reproduce experimental data, it is worth rigorously investigating the renormalization of the NLO contribution to the magnetic moment. While speculative at this point, it is possible that such an evaluation could indicate the need to promote a CT operator to higher orders and explain the lack of convergence in the power counting observed in this study and the studies in Refs. [10, 11]. Further, Ref. [14] previously argued on the basis of the wave function anomalous dimension that the CT operator contributing to the magnetic moment should have a power counting that places it roughly between N2LO and NLO, consistent with the observed enhancement; however, such an effect could also arise simply due to cutoffs that are too soft causing an uncontrolled sensitivity to the short-range physics [15]. Either case has important implications for the convergence of the power counting and for future studies of electromagnetic observables. As such, studies of the renormalization of simple two-body observables are presently being pursued to further understand these observations and to determine the nature of this enhancement.

To better compare with experimental data, spurious contamination should be removed from the VMC wave functions through GFMC propagation. Work along this lines is presently underway using model Ia*.

6.3 Two-Body Magnetic Densities

To further investigate the cutoff dependence in the above calculations and examine the spacial distribution of two-body contributions, two-body magnetic densities ρ_M^{2b} are also computed. The density is defined such that, for interparticle spacings r_{ij},

$$\mu^{2b} = \int dr_{ij} 4\pi r_{ij}^2 \rho_M^{2b}(r_{ij}) \,, \tag{6.4}$$

for all two-body contributions to the total magnetic moment. The computation of the densities is similar in spirit to Eq. 4.6, however we must now extract the low-momentum transfer limit of the relevant matrix element. As in Sect. 4.1.2, CT contributions require two nucleons close together and thus will provide a purely short-range contribution, while long-range ($r_{ij} \gtrsim 1$ fm) tails appear in OPE-like densities.

For the IS nuclei, all but three contributions vanish due to symmetry and it is therefore the simplest case to discuss. Figure 6.2 displays the contributions to ρ_M^{2b} in ^{6}Li and ^{10}B from the IS current computed in model IIb*. As was previously observed in Ref. [16], the N3LO(OPE) IS contribution displays a node that suppresses the overall contribution from the matrix element. Schematically, the current is given as,

$$\mathbf{j}_{\mathrm{OPE(IS)}}^{\mathrm{N3LO}} = I_1(\mu_{ij})\mathcal{O}_{ij}^{(1)} + I_2(\mu_{ij})\mathcal{O}_{ij}^{(2)} \,, \tag{6.5}$$

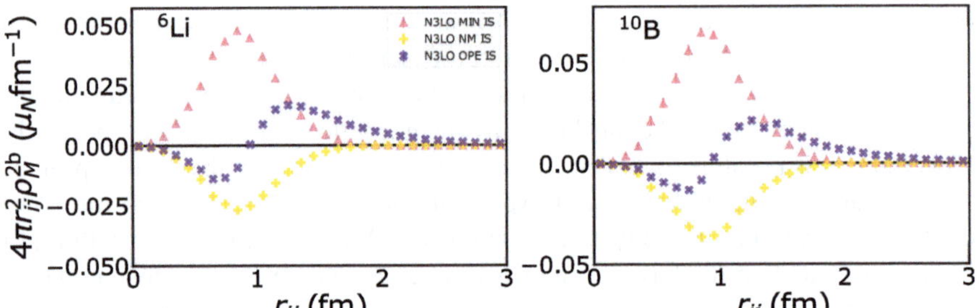

Fig. 6.2 Two-body VMC magnetic densities for the ^{6}Li and ^{10}B ground states computed with the NV2+3-IIb* interaction and currents as a function of interparticle spacing r_{ij}. Statistical uncertainties are included but smaller than the markers representing the central value at a given r_{ij}. Figure reproduced with permission from Ref. [9]

where $\mu_{ij} = m_\pi r_{ij}$ and $\mathcal{O}_{ij}^{(1)}$ and $\mathcal{O}_{ij}^{(2)}$ indicate different operator structures. The correlation functions have the forms

$$I_1(x) = C(1+x)\frac{e^{-x}}{x^3} \tag{6.6}$$

$$I_2(x) = -C(3+3x+x^2)\frac{e^{-x}}{x^3}, \tag{6.7}$$

where C denotes the combination of LECs governing their strength (see Eq. A.74 for the exact expression). The opposite signs of the correlations lead to the cancellation that causes the node and thus the suppression of this contribution. This, combined with the cancellation of the N3LO(MIN) and N3LO(NM) CT contributions, leads to a small contribution from IS two-body currents to the total magnetic moment.

In Fig. 6.3, the IV two-body magnetic densities are displayed for $A = 3,\ 7,$ and 8 nuclei. Immediately, one notes that the two-body currents in mirror systems appear to be a reflection of each other, having strikingly similar shapes with opposite sign. Another notable feature is the prominent N3LO(NM) peak, which appears in between the peaks for the NLO(OPE) and N2LO(Δ) term. The other N3LO curves are not as significantly enhanced, with peaks below the N2LO(Δ) contribution. In the N3LO(OPE) IV term, there are two different operator structures present and their strengths are governed by two different LECs—d_8' or d_{21}' (see Eqs. A.85 and A.86) [16, 17]. These two terms end up producing magnetic densities with opposite signs that cancel out to quench the integrated N3LO(OPE) IV contribution. This is particularly striking for model Ia*, where the N3LO(OPE) contribution nearly vanishes for all nuclei.

In $A = 9$, one again observes that for mirror nuclei, the two-body transition densities have a similar shape and differ only in the sign of the contribution. As in Fig. 6.3, the N3LO(NM) transition density is enhanced and its peak falls between the NLO and N2LO contributions. One again sees the interplay between the two N3LO OPE operator structures that cancel each other to quench the overall contribution of this term to the magnetic moment. In the long-range transition densities, for the $T = 1/2$ nuclei, there is a node present in the tail. Similar to what happens for the contribution from the N3LO(OPE) IS current, there is an interplay of different radial structures in the current operators that can change the sign of the contribution (see Eq. A.76) (Fig. 6.4).

As expected from the cluster structure in the $T = 1/2$ nuclei, the NLO OPE contribution is suppressed compared to its value for both the $T = 3/2$ nuclei with $A = 9$ and to the $A = 3,\ 7,$ and 8 nuclei. The NLO density for the $T = 3/2$ nuclei is particularly enhanced around its peak at $r_{ij} \approx 1$ fm. This is likely due to the [432] structure allowing for the two valence nucleons to be spatially symmetric with a non-zero pair wave function at shorter distances. The reduction of the CT contributions in $T = 1/2$ is not explained by the pion exchange arguments, but is still linked to the [441] cluster structure. As an example, note that the NL3O NM operator (Eqs. A.87 and A.88) is schematically given by [16],

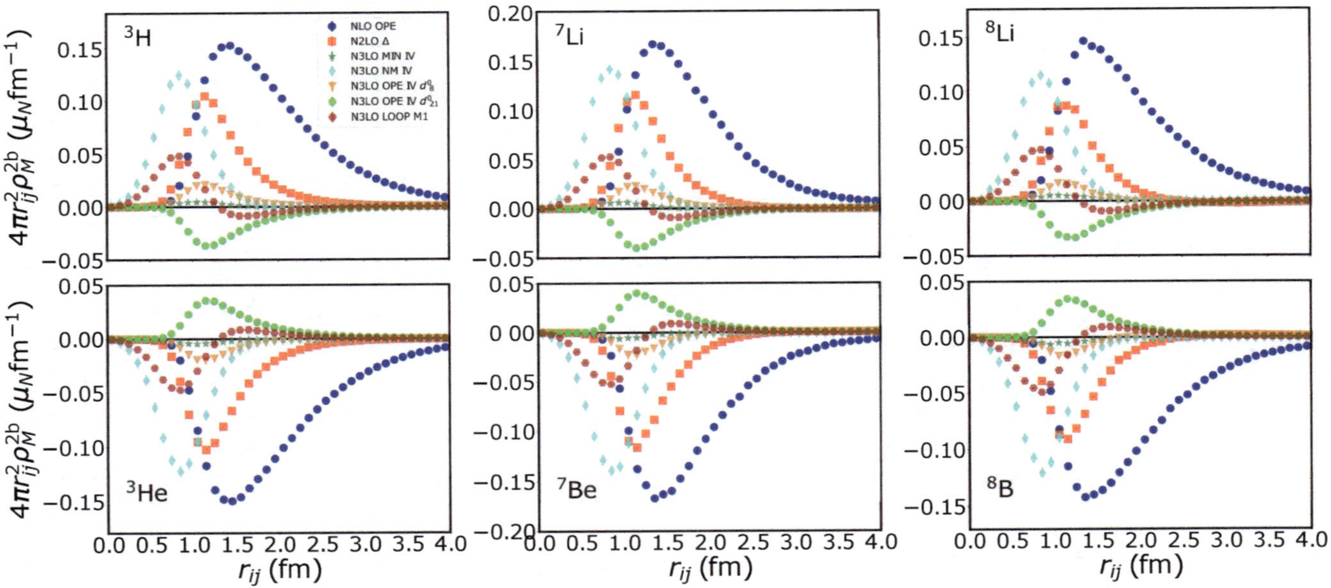

Fig. 6.3 Two-body VMC IV magnetic densities for the $A = 3, 7,$ and 8 ground states computed with the NV2+3-IIb* interaction and currents as a function of interparticle spacing r_{ij}. Statistical uncertainties are included but smaller than the markers representing the central value at a given r_{ij}. Figure reproduced with permission from Ref. [9]

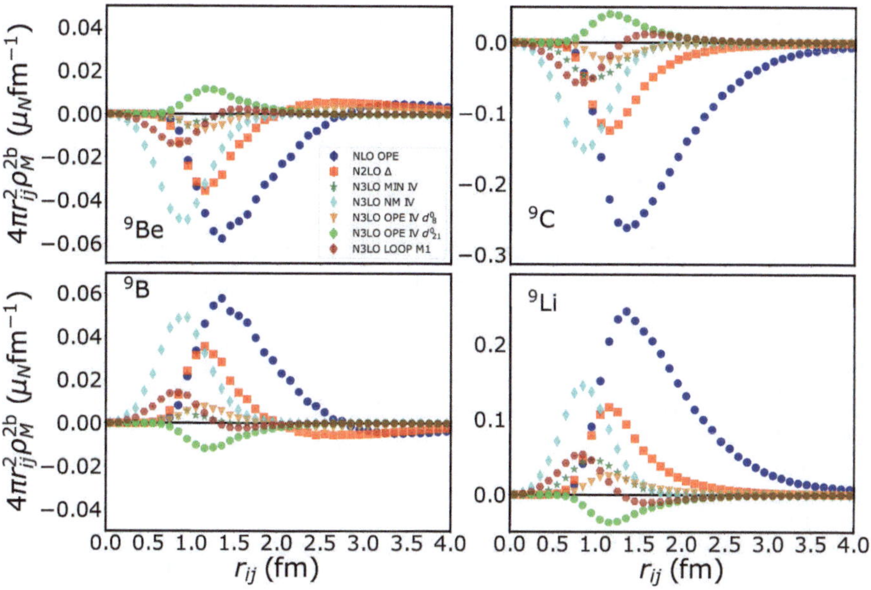

Fig. 6.4 Two-body VMC IV magnetic densities for the $A = 9$ ground states computed with the NV2+3-IIb* interaction and currents as a function of interparticle spacing r_{ij}. Statistical uncertainties are included but smaller than the markers representing the central value at a given r_{ij}. Figure reproduced with permission from Ref. [9]

$$
\begin{aligned}
\mathbf{j} = & -ie^{i\mathbf{q}\cdot\mathbf{R}_{ij}} C_{R_S}(r_{ij}) \\
& \times \left[\widetilde{C}(\boldsymbol{\sigma}_i + \boldsymbol{\sigma}_j) + \overline{C}(\tau_{i,z} - \tau_{j,z})(\boldsymbol{\sigma}_i - \boldsymbol{\sigma}_j)\right] \times \mathbf{q}\,,
\end{aligned}
\tag{6.8}
$$

where $\widetilde{C}$ and $\overline{C}$ are standing in for different combinations of the pion mass and LECs. At short distances, the IV pairs tend to be $ST = 01$ due to the suppression of P-waves at short distances. For these pairs, the expectations value of $(\boldsymbol{\sigma}_i + \boldsymbol{\sigma}_j)_y$ and $(\tau_{i,z} - \tau_{j,z})$ vanish. As the $A, T = 9, 1/2$ nuclei do not significantly admix other symmetries in the ground state, this leads to the overall suppression of the CT densities at short distances.

Similar arguments may also explain why the N3LO(NM) density in the $A, T = 9, 3/2$ nuclei is relatively unchanged going from $A = 3$ and 7 while the N3LO(MIN) contribution is enhanced. Since the additional two nucleons in $A, T = 9, 3/2$ systems are spatially antisymmetric from those in the in the [43] cluster, the only significant new CT contribution should come from the correlation between themselves. That contribution to the N3LO(NM) current, which is purely proportional to $(\tau_{i,z} - \tau_{j,z})$, vanishes for two neutrons or two protons. Instead, the N3LO(MIN) current contains additional term proportional to $(\boldsymbol{\tau}_i \times \boldsymbol{\tau}_j)_z$ and is thus impacted by the additional contribution from the valence pair.

To further understand the behavior of ρ_M^{2b}, it is interesting to compare the shapes of the transition densities. That is, one can look at the magnetic densities scaled by

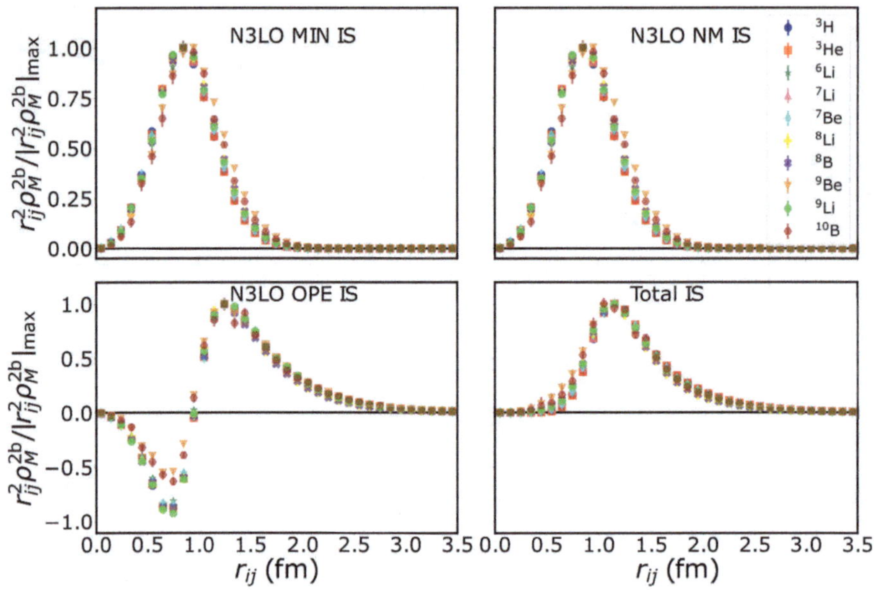

Fig. 6.5 Scaled two-body VMC IS magnetic densities for different nuclei computed with model IIb* as a function of interparticle spacing r_{ij}. Different colors represent different nuclei. Figure reproduced with permission from Ref. [9]

the maximum value of $|r_{ij}\rho_M^{2b}|$. In Figs. 6.5 and 6.6, I compare the scaled IS and IV transitions densities, respectively. The scaled IS magnetic densities display a striking, nearly universal behavior across r_{ij}. On the other hand, the scaled IV densities are universal at short range and display a tail which depends on the particular nucleus understudy. In Ref. [18] and reported on in Sect. 4.1, a similar behavior was observed for axial transition densities in a study of Gamow-Teller β-decay matrix elements. The explanation for the universal behavior of the magnetic densities follows a similar logic. At short distances ($r_{ij} \lesssim \hbar c/m_\pi$), there are very few pairs in relative P-waves due to the repulsive nature of the nuclear force in odd partial waves. Thus, IS pairs at short distances tend to be predominantly $ST = 10$ and IV pairs $ST = 01$. Both the $ST = 01$ and $ST = 10$ pair distributions in nuclei have similar shapes and differ only by an overall scaling factor [19, 20]. Therefore, this universal behavior is a result of the short-range correlations in nuclei that are driven by the tensor component of the nuclear force [21]. At larger distances, since there is less suppression of P-wave pairs and because the $ST = 00$ and 11 pair distributions do not scale, the two-body magnetic densities can lose their universal scaling; however, because $ST = 00$ pairs are formed rather infrequently in light systems [13, 22], the long-range tail of the IS OPE term does not strongly deviate from universal behavior. Instead, it is possible to form a large number of $ST = 11$ pairs in light systems which spoils universality for the total IV term. This, the near universal behavior of the total IS contribution, the short range universality of the IV term, and the long-range nucleus dependence

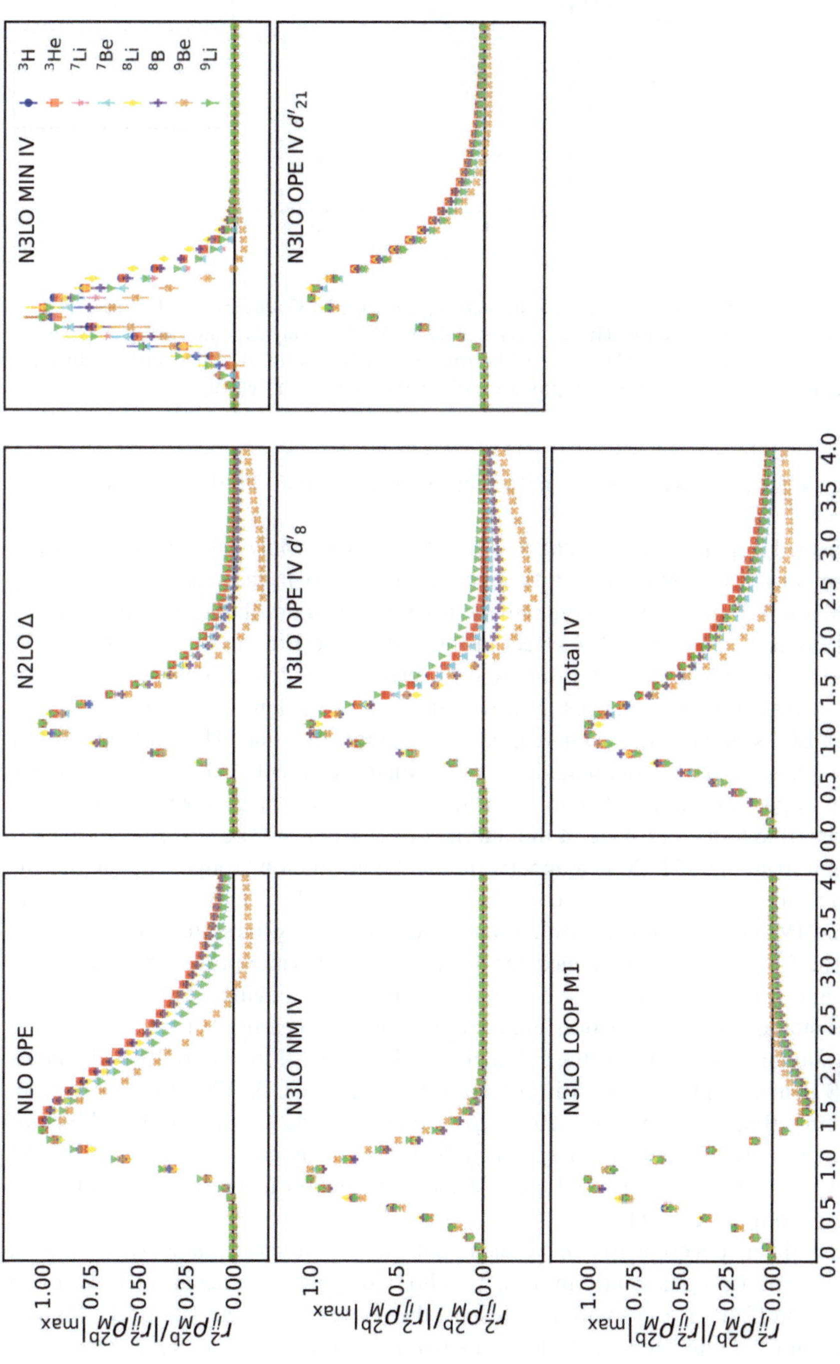

Fig. 6.6 Scaled two-body VMC IV magnetic densities for different nuclei computed with model IIb* as a function of interparticle spacing r_{ij}. Different colors represent different nuclei. Figure reproduced with permission from Ref. [9]

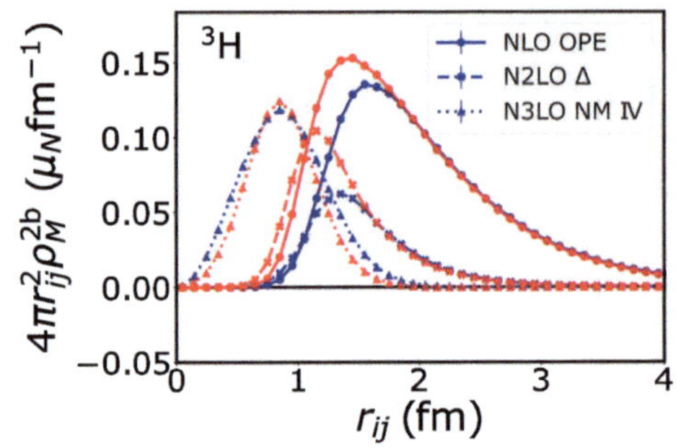

Fig. 6.7 Two-body VMC magnetic densities arising from the NLO (circles by solid lines), N2LO-Δ (squares connected by dashed lines), and the N3LO(NM) IV (triangles connected by dotted lines) contributoins for ^{3}H computed with model Ia* (blue symbols) and IIb* (red symbols) as a function of interparticle spacing r_{ij}. Figure reproduced with permission from Ref. [9]

of the IV term are well understood in terms of the formation of pairs inside of the nucleus.

Up to this point, I have analyzed the densities for only one model; however, it is important to assess the cutoff dependence present in the integrated two-body contributions. To that end, I compare the dominant IV contributions to the ^{3}H magnetic moment in Fig. 6.7. The effect of the softer cutoff in the Ia* regulator on the NLO(OPE) and N2LO(Δ) currents is, unsurprisingly, to reduce the two-body magnetic density at short distances; however, the tail of the density, which depends on known LECs in the NV2, is unaffected by the cutoff function. Thus, taking a softer cutoff quenches the overall magnitude of the matrix element. This has a particularly strong impact on the N2LO current, which is of shorter-range than the NLO contribution due to the presence of the intermediate Δ-isobar. The soft cutoff removes strength from the N2LO(Δ) magnetic moment contribution around its peak, resulting in a large quenching of the contribution in model Ia* relative to model IIb*. The N3LO(NM) contribution, however, is enhanced when using the softer cutoff of model Ia*. The CT density is broadened for the softer cutoff while preserving the overall magnitude. This effect is likely due to a combination of having a broader Gaussian representation of the δ-function and a larger density of nearby IV pairs.

A similar broadening of the CT densities is observed in Fig. 6.8 which shows the IS densities plotted for ^{6}Li. For the long-range N3LO(OPE) IS contribution, the correlation functions $I_1(\mu)$ and $I_2(\mu)$ are particularly sensitive to the choice of regulator, as was previously observed in Ref. [16]. This can move the node in the density and change the overall sign of the contribution to the magnetic moment coming from this current.

From the model comparison, it is clear that altering the short-range dynamics with different cutoffs significantly impacts two-body magnetic moments. Clearly, if one takes a cutoff that is too small, it will spoil the convergence behavior of the higher-order contributions due to significant quenching of the NLO and N2LO terms that gets compensated by a larger N3LO(NM) IV contribution. This model dependence is

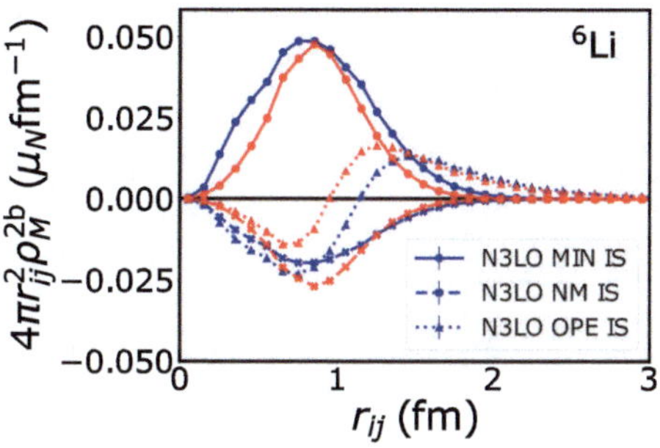

Fig. 6.8 Two-body VMC magnetic densities arising from the N3LO(MIN) IS (circles by solid lines), N3LO(NM) IS (squares connected by dashed lines), and the N3LO(OPE) IS (triangles connected by dotted lines) contributions for ^{6}Li computed with model Ia⋆ (blue symbols) and IIb⋆ (red symbols) as a function of interparticle spacing r_{ij}. Figure reproduced with permission from Ref. [9]

obscured if one looks only at the total magnetic moment. Given that there is a strong model dependence of the two-body contributions and the particular importance of the IV current in predicting magnetic structure, it is crucial to investigate this cutoff dependence more rigorously in the future.

6.4 Magnetic Form Factors

Up to this point in the chapter, I have focused on magnetic moments, which are a low momentum transfer observable; however, from the elastic scattering of electrons on the nucleus, we have information on the magnetic structure of nuclei over a broad range of momentum transfers. By comparing to magnetic form factors of nuclei extracted from experimental data, it is possible to evaluate the zero energy transfer behavior of the model at several momentum transfers. This was done in Refs. [9, 23], the results of which are summarized in this section.

The magnetic form factor in Eq. 6.1 was computed using the same interactions and currents as were used for the magnetic moments. In Figs. 6.9 and 6.10, I present the results for all the nuclei showing the contributions of the currents order-by-order for the IIb⋆ interaction compared to data where available. The VMC calculations capture very well the shape of the experimental data, though the magnitude and widths of diffraction peaks are not always reproduced perfectly. In addition to the form factors for which there are data, the results also include predicted magnetic form factors of radioactive isotopes. While there are presently no data to compare for the radioactive systems, novel techniques have recently made it possible to study electron scattering from unstable nuclei [24] and advances could make such comparisons possible in the future. In Fig. 6.11, the multipole contributions to the magnetic form factors for

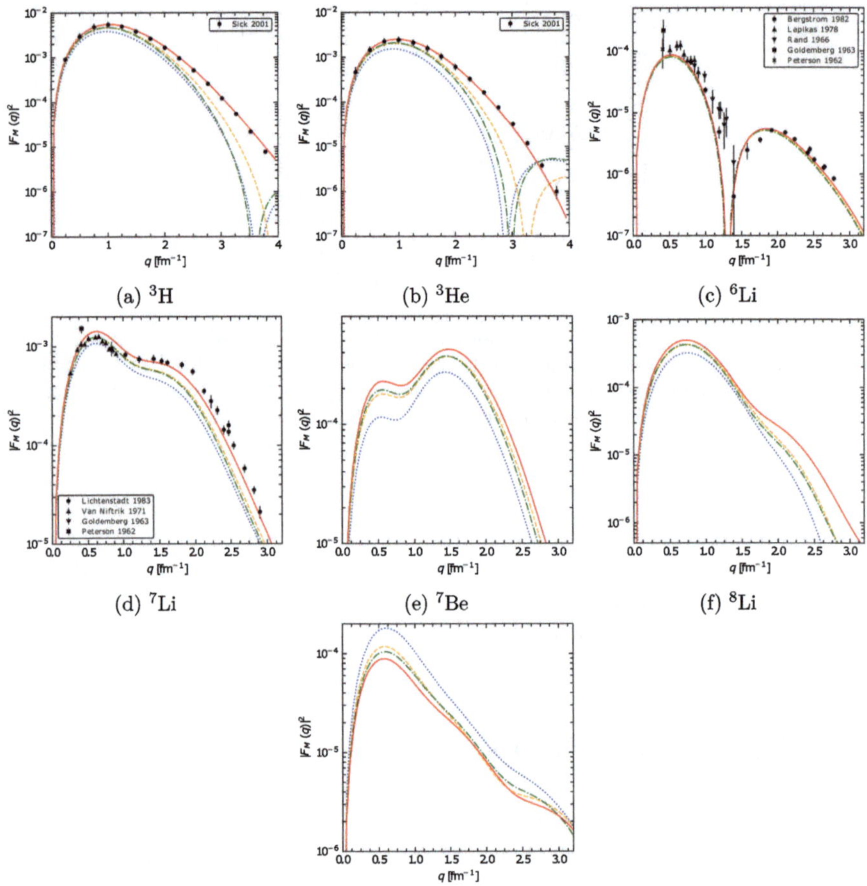

(a) ^{3}H (b) ^{3}He (c) ^{6}Li

(d) ^{7}Li (e) ^{7}Be (f) ^{8}Li

(g) ^{8}B

Fig. 6.9 Order-by-order VMC magnetic form factors computed with the NV2+3IIb* potential and consistent electromagnetic current for $A = 3,\ 6,\ 7,$ and 8 nuclei. The contributions of the currents up to LO (dotted blue), NLO (dashed yellow), N2LO (dashed dotted green) and N3LO (full red) in the chiral expansion are shown. The calculations are compared to the available world experimental data. Figure reproduced with permission from Ref. [9]

the nuclei with $J > 1$ computed with the currents at N3LO and the IIb* are shown. Similar results are obtained for the other interaction and current models considered.

The VMC calculation obtains good agreement with the experimental form factors of ^{3}H and ^{3}He [25, 26], as can be seen in panels (a) and (b) of Fig. 6.9 for the IIb* interactions. Clearly, without two-body currents, the model would not be able to predict the data at $q \sim 3.5$ fm^{-1}. The diffraction minimum is filled is by the NM and OPE terms at N3LO, which are fit to data. A similar observation was made previously in Ref. [10].

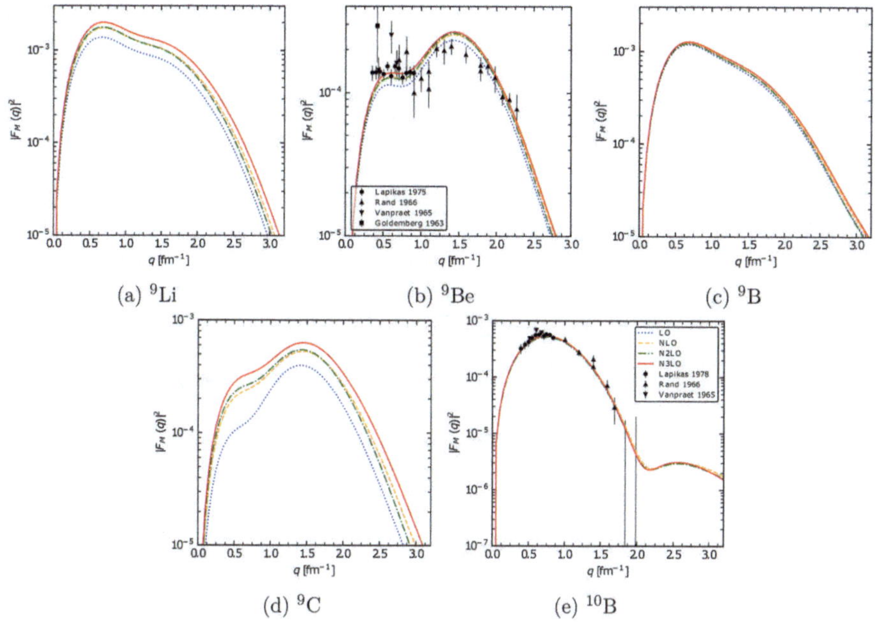

Fig. 6.10 Order-by-order VMC magnetic form factors computed with the NV2+3IIb* potential and consistent electromagnetic current for $A = 9$, and 10 nuclei. The contributions of the currents up to LO (dotted blue), NLO (dashed yellow), N2LO (dashed dotted green) and N3LO (full red) in the chiral expansion are shown. The calculations are compared to the available world experimental data. Figure reproduced with permission from Ref. [9]

Using the VMC approach, I analyzed the magnetic form factors for heavier systems. The first such system with experimental data is ^{6}Li and the comparison with theory is shown in Fig. 6.9c. Unsurprisingly, two-body currents do not play a large role in this form factor, as the ^{6}Li ground state is purely IS and the first subleading contribution comes at N3LO. Reasonable agreement with the data from Refs. [27–31] is obtained, with the two-peaked shape being reproduced; however, the diffraction minimum is shifted slightly to lower momentum transfer, due to the overly narrow first peak. Further, the strength of the second peak makes the calculation overshoot the data.

For $A = 7$, we once again see that many-body currents play an important role due to the contribution from the IV currents beginning at NLO plus the fitted IV terms at N3LO that made rather large contributions. In the case of ^{7}Li, there is available data from Refs. [27, 28, 32, 33]. While the overall shape and magnitude of the data are well reproduced in Fig. 6.9d, there is not exact agreement between the VMC calculation and the measurement. At low q, the VMC result up to N3LO overshoots the data, while at larger q, there is missing strength. The two body contribution for the first peak is roughly 25%, while nearly half the strength in the tail comes from

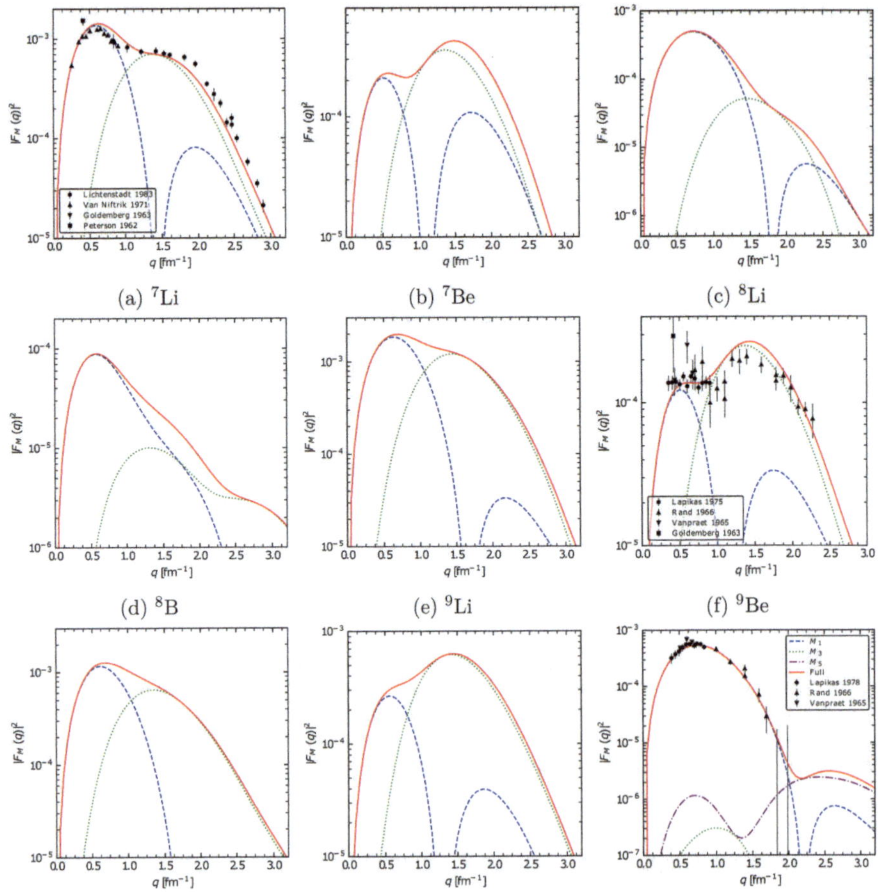

(a) ^{7}Li (b) ^{7}Be (c) ^{8}Li
(d) ^{8}B (e) ^{9}Li (f) ^{9}Be
(g) ^{9}B (h) ^{9}C (i) ^{10}B

Fig. 6.11 Breakdown of the multipole contributions to the form factors computed in VMC with the NV2+3-IIb* potential and consistent electromagnetic current at N3LO for nuclei with $J > 1$. The contribution is broken down according to that coming from the multipoles M_1 (dashed blue line), M_3 (dotted green), and M_5 (dashed dotted purple). The red line represents the summed result and comparison is made to the available world data. Figure reproduced with permission from Ref. [9]

two-body currents. Large corrections from meson exchange currents were already observed in phenomenological approaches, as noted in Ref. [2].

In Fig. 6.9e, one sees that the contribution from two-body currents for ^{7}Be is even larger for the two-peaks present in the form factor than for its mirror. While no data exist, the possibility of using electron storage rings at rare isotope beam facilities [24] could make this a candidate to measure and better understand two-body currents in the future, if the technique becomes sufficiently advanced. It is also interesting to note the difference in the dominant peaks, with the first peak associated with the M_1 multipole being suppressed relative to the second peak associated with

M_3 for ^{7}Be (see Fig. 6.11b). In the mirror, the opposite is true (Fig. 6.11a). This may be understood in terms of the different structures of the two systems. The leading contribution to M_1 in ^{7}Be approximately comes from coupling between an external field to a neutron, while in ^{7}Li, the leading contribution is approximately from the coupling to a proton. Because the gyromagnetic ratio of the the neutron magnetic moment is smaller than that of the proton, the first peak of ^{7}Be is quenched relative to ^{7}Li.

For $A = 8$, there are no experimental data for the magnetic form factors; however, we observe again the large contribution from two-body currents (Fig. 6.9, panels (f) and (g)). This is particularly striking for ^{8}B, where two-body currents reduce the strength of the form factor for all q. It is also worth noting the difference in the structure imprinted on the calculations, with ^{8}B only having a single peak at low q while ^{8}Li exhibits a two-peak structure. Should advancements in electron scattering on radioisotopes progress enough to measure these form factors, they would also provide interesting cases to better understand two-body currents.

The $A = 9$ systems once again reflect the interplay between structure and many-body currents. For the $A, T = 9, 3/2$ nuclei, two-body currents play a large role, as they did for the magnetic moment. ^{9}C is a particularly striking, with two-body currents making up most of the strength in the first peak (Fig. 6.10d). In Fig. 6.10a, one sees that the mirror system ^{9}Li has a strong two-body effect, but not as pronounced as that of ^{9}C. As with $A = 7$, there is similar inversion of the dominant multipole from M_1 to M_3 going from ^{9}Li to ^{9}C (see in Fig. 6.11 panels (e) and (h)) arising from the cluster structure of these systems.

Moving to the $A, T = 9, 1/2$ systems, with form factors plotted in Fig. 6.10 panels (b) and (c), have two-body currents suppressed due to their $2\alpha + N$ cluster structures resulting in weak OPE binding felt by the valence nucleon and the formation of short-range pairs between the valence and nucleons inside of the 2α structure quelled by the Pauli principle. The mirror systems again reflect the cluster structure, with ^{9}B having a larger M_1 peak than M_3 peak and the opposite for ^{9}Be (see Fig. 6.11 panels (f) and (g)). The only available data for these systems is for ^{9}Be [28, 29, 34, 35]. The data suggest a plateau at low q and a second peak near $q \sim 1.7\ \text{fm}^{-1}$. While the VMC calculation reproduces the plateau, the two-body currents provide too much strength and overshoot the peak. In the review of Ref. [2], calculations presented therein could reproduce the second peak with phenomenological models but underestimated the plateau. Incidentally, previous projected Hartree-Fock calculation shows a result very similar to the VMC, where the secondary peak is overestimated even if no two-body currents are included [36].

The final case studied was ^{10}B. Being an IS nucleus, the two-body current contribution is suppressed. In this case, the VMC calculation provides excellent agreement with the data in Refs. [28–30, 35, 37], which cover the range $0.5 \lesssim q \lesssim 2.0\ \text{fm}^{-1}$. This region is dominated by the contribution from M_1. No data are present in the second peak region, which is dominated by the M_5 multipole contribution, so data in the large q range could provide an assessment of the quality of higher-order multipoles extracted from VMC calculations of matrix elements.

References

1. S. Pastore, Steven C. Pieper, R. Schiavilla, R.B. Wiringa, Quantum Monte Carlo calculations of electromagnetic moments and transitions in A $\leq$ 9 nuclei with meson-exchange currents derived from chiral effective field theory. Phys. Rev. **C87**(3), 035503 (2013). https://doi.org/10.1103/PhysRevC.87.035503
2. T.W. Donnelly, I. Sick, Elastic magnetic electron scattering from nuclei. Rev. Mod. Phys. **56**, 461–566 (1984). https://doi.org/10.1103/RevModPhys.56.461
3. J. Carlson, S. Gandolfi, F. Pederiva, Steven C. Pieper, R. Schiavilla, K.E. Schmidt, R.B. Wiringa, Quantum Monte Carlo methods for nuclear physics. Rev. Mod. Phys. **87**, 1067 (2015). https://doi.org/10.1103/RevModPhys.87.1067
4. G.B. King, A. Baroni, V. Cirigliano, S. Gandolfi, L. Hayen, E. Mereghetti, S. Pastore, M. Piarulli, Ab initio calculation of the β-decay spectrum of He6. Phys. Rev. C **107**(1), 015503 (2023). https://doi.org/10.1103/PhysRevC.107.015503
5. J.E. Purcell, J.H. Kelley, E. Kwan, C.G. Sheu, H.R. Weller, Energy levels of light nuclei A = 3. Nucl. Phys. A **848**, 1–74 (2010). https://doi.org/10.1016/j.nuclphysa.2010.08.012
6. D.R. Tilley, C.M. Cheves, J.L. Godwin, G.M. Hale, H.M. Hofmann, J.H. Kelley, C.G. Sheu, H.R. Weller, Energy levels of light nuclei A=5, A=6, A=7. Nucl. Phys. A **708**, 3–163 (2002). https://doi.org/10.1016/S0375-9474(02)00597-3
7. D.R. Tilley, J.H. Kelley, J.L. Godwin, D.J. Millener, J.E. Purcell, C.G. Sheu, H.R. Weller, Energy levels of light nuclei A=8,9,10. Nucl. Phys. A **745**, 155–362 (2004). https://doi.org/10.1016/j.nuclphysa.2004.09.059
8. D. Borremans et al., New measurement and reevaluation of the nuclear magnetic andquadrupole moments of Li-8 and Li-9. Phys. Rev. C **72**, 044309 (2005). https://doi.org/10.1103/PhysRevC.72.059902. [Erratum: Phys. Rev. C 72, 059902 (2005)]
9. G. Chambers-Wall, A. Gnech, G.B. King, S. Pastore, M. Piarulli, R. Schiavilla, R.B. Wiringa, Magnetic structure of A$\leq$10 nuclei using the Norfolk nuclear models with quantum Monte Carlo methods. Phys. Rev. C **110**(5), 054316 (2024). https://doi.org/10.1103/PhysRevC.110.054316
10. A. Gnech, R. Schiavilla, Magnetic structure of few-nucleon systems at high momentum transfers in a chiral effective field theory approach. Phys. Rev. C **106**(4), 044001 (2022). https://doi.org/10.1103/PhysRevC.106.044001
11. J.D. Martin, S.J. Novario, D. Lonardoni, J. Carlson, S. Gandolfi, I. Tews, Auxiliary field diffusion Monte Carlo calculations of magnetic moments of light nuclei with chiral effective field theory interactions. Phys. Rev. C **108**(3), L031304 (2023). https://doi.org/10.1103/PhysRevC.108.L031304
12. E. Tiesinga, P.J. Mohr, D.B. Newell, B.N. Taylor, CODATA recommended values of the fundamental physical constants: 2018*. Rev. Mod. Phys. **93**(2), 025010 (2021). https://doi.org/10.1103/RevModPhys.93.025010
13. R.B. Wiringa, Pair counting, pion-exchange forces, and the structure of light nuclei. Phys. Rev. C **73**, 034317 (2006). https://doi.org/10.1103/PhysRevC.73.034317
14. M. Pavón Valderrama, D.R. Phillips, Power Counting of Contact-Range Currents in Effective Field Theory. Phys. Rev. Lett. **114**(8), 082502 (2015). https://doi.org/10.1103/PhysRevLett.114.082502
15. H.W. Hammer, S. König, U. van Kolck, Nuclear effective field theory: status and perspectives. Rev. Mod. Phys. **92**(2), 025004 (2020). https://doi.org/10.1103/RevModPhys.92.025004
16. R. Schiavilla, A. Baroni, S. Pastore, M. Piarulli, L. Girlanda, A. Kievsky, A. Lovato, L.E. Marcucci, Steven C. Pieper, M. Viviani, R.B. Wiringa, Local chiral interactions and magnetic structure of few-nucleon systems. Phys. Rev. C **99**, 034005 (2019). https://doi.org/10.1103/PhysRevC.99.034005
17. S. Pastore, L. Girlanda, R. Schiavilla, M. Viviani, R.B. Wiringa, Electromagnetic Currents and Magnetic Moments in (chi)EFT. Phys. Rev. C **80**, 034004 (2009). https://doi.org/10.1103/PhysRevC.80.034004

18. G.B. King, L. Andreoli, S. Pastore, M. Piarulli, R. Schiavilla, R.B. Wiringa, J. Carlson, S. Gandolfi, Chiral effective field theory calculations of weak transitions in light nuclei. Phys. Rev. C **102**, 025501 (2020). https://doi.org/10.1103/PhysRevC.102.025501
19. J.L. Forest, V.R. Pandharipande, S.C. Pieper, R.B. Wiringa, R. Schiavilla, A. Arriaga, Femtometer toroidal structures in nuclei. Phys. Rev. C **54**, 646–667 (1996). https://doi.org/10.1103/PhysRevC.54.646
20. R. Cruz-Torres, D. Lonardoni, R. Weiss, N. Barnea, D.W. Higinbotham, E. Piasetzky, A. Schmidt, L.B. Weinstein, R.B. Wiringa, O. Hen, Many-body factorization and position-momentum equivalence of nuclear short-range correlations. Nature Phys. **17**(3), 306–310 (2021). https://doi.org/10.1038/s41567-020-01053-7
21. R. Schiavilla, R.B. Wiringa, S.C. Pieper, J. Carlson, Tensor Forces and the Ground-State Structure of Nuclei. Phys. Rev. Lett. **98**, 132501 (2007). https://doi.org/10.1103/PhysRevLett.98.132501
22. M. Piarulli, S. Pastore, R.B. Wiringa, S. Brusilow, R. Lim, Densities and momentum distributions in A≤12 nuclei from chiral effective field theory interactions. Phys. Rev. C **107**(1), 014314 (2023). https://doi.org/10.1103/PhysRevC.107.014314
23. G. Chambers-Wall, A. Gnech, G.B. King, S. Pastore, M. Piarulli, R. Schiavilla, R.B. Wiringa, Quantum Monte Carlo Calculations of Magnetic Form Factors in Light Nuclei. Phys. Rev. Lett. **133**(21), 212501 (2024). https://doi.org/10.1103/PhysRevLett.133.212501
24. K. Tsukada, Y. Abe, A. Enokizono, T. Goke, M. Hara, Y. Honda, T. Hori, S. Ichikawa, Y. Ito, K. Kurita, C. Legris, Y. Maehara, T. Ohnishi, R. Ogawara, T. Suda, T. Tamae, M. Wakasugi, M. Watanabe, H. Wauke, First observation of electron scattering from online-produced radioactive target. Phys. Rev. Lett. **131**, 092502 (Aug2023), https://doi.org/10.1103/PhysRevLett.131.092502. URL https://link.aps.org/doi/10.1103/PhysRevLett.131.092502
25. A. Amroun et al., H-3 and He-3 electromagnetic form-factors. Nucl. Phys. A **579**, 596–626 (1994). https://doi.org/10.1016/0375-9474(94)90925-3
26. I. Sick. Elastic electron scattering from light nuclei. *Progress in Particle and Nuclear Physics*, 47(1):245–318, 2001. ISSN 0146-6410. https://doi.org/10.1016/S0146-6410(01)00156-9. https://www.sciencedirect.com/science/article/pii/S0146641001001569
27. G.A. Peterson, Magnetic form-factors by 180° electron scattering. Phys. Lett. **2**(3), 162–163 (1962). ISSN 0031-9163. https://doi.org/10.1016/0031-9163(62)90226-3. https://www.sciencedirect.com/science/article/pii/0031916362902263
28. J. Goldemberg, Y. Torizuka, Magnetic elastic scattering of electrons by light nuclei. Phys. Rev. **129**, 312–315 (Jan1963), https://doi.org/10.1103/PhysRev.129.312. URL https://link.aps.org/doi/10.1103/PhysRev.129.312
29. R.E. Rand, R. Frosch, M.R. Yearian, Elastic electron scattering from the magnetic multipole distributions of li^6, li^7, be^9, b^{10}, b^{11}, and n^{14}. Phys. Rev. **144**, 859–873 (Apr1966), https://doi.org/10.1103/PhysRev.144.859. URL https://link.aps.org/doi/10.1103/PhysRev.144.859
30. L. Lapikás, Proceedings of the Conference on Modern Trends in Elastic Electron Scattering. C. deVries, NIKHEF-K (Amsterdam, 1978)
31. J.C. Bergstrom, S.B. Kowalski, R. Neuhausen, Elastic magnetic form factor of ^{6}Li. Phys. Rev. C **25**, 1156–1167 (Mar1982), https://doi.org/10.1103/PhysRevC.25.1156
32. G.J.C. Van Niftrik, L. Lapikás, H. De Vries, G. Box, Magnetization distribution of the 7li nucleus as obtained from electron scattering through 180°. the electric quadrupole moment of 7li. Nuclear Phys. A **174**(1), 173–192 (1971). ISSN 0375-9474. https://doi.org/10.1016/0375-9474(71)91011-6. https://www.sciencedirect.com/science/article/pii/0375947471910116
33. J. Lichtenstadt, J. Alster, M.A. Moinester, J. Dubach, R.S. Hicks, G.A. Peterson, S. Kowalski, The low lying levels of 7li studied by electron scattering. Phys. Lett. B **121**(6), 377–380 (1983). ISSN 0370-2693. https://doi.org/10.1016/0370-2693(83)91181-4. https://www.sciencedirect.com/science/article/pii/0370269383911814
34. L. Lapikás, G. Box, H. De Vries, Magnetic elastic electron scattering from 3be and 13c. Nuclear Phys. A **253**(2), 324–354 (1975). ISSN 0375-9474. https://doi.org/10.1016/0375-9474(75)90485-6. https://www.sciencedirect.com/science/article/pii/0375947475904856

35. G.J. Vanpraet and P. Kossanyi-Demay, Magnetic elastic electron scattering from ^{9}be, ^{10}b and ^{11}b. Il Nuovo Cimento A **63**(1), 388–390 (1965). ISSN 1826-9869. https://doi.org/10.1007/BF02898849
36. M. Bouten, M.C. Bouten, H. Depuydt, Elastic electron scattering from nuclear magnetic moments in 9 Be and projected Hartree-Fock wave functions. Nucl. Phys. A **131**, 385–392 (1969). https://doi.org/10.1016/0375-9474(69)90542-9
37. J. Goldemberg, D.B. Isabelle, T. Stovall, D. Vinciguerra, A. Bottino, Investigation of boron 10 and boron 11 by elastic electron scattering at 180°. Phys. Lett. **16**(2), 141–143 (1965). ISSN 0031-9163. https://doi.org/10.1016/0031-9163(65)90159-9. https://www.sciencedirect.com/science/article/pii/0031916365901599

Chapter 7
Conclusion and Outlook

Abstract In this chapter, I provide concluding remarks on the thesis. First, I summarize the results, and how the many-body correlations and two-body currents played a role in the calculations detailed in earlier chapters. I also discuss potential extensions of this work, and how they can impact searches of beyond Standard Model physics.

In this thesis, I compute electroweak observables over a wide range of energy and momentum transfer using quantum Monte Carlo many-body techniques and the microscopic theoretical framework based on the NV2+3 models and the associated one- and two-body electroweak currents. I focused on the role of many-body dynamics in these processes; namely, I studied in detail the effects of many-nucleon correlations and currents, as well as their role in achieving agreement with the available experimental data. The novel contribution of two-body electroweak operator densities helped to strengthen these analyses and gave insight into the dynamics that underlie many-body contributions to electroweak processes. These densities highlighted the universality of short-range physics and the differences between nuclear models. Finally, informed the observed model dependencies, an *ad hoc* procedure for estimating uncertainties on predicted quantities was developed. This microscopic comprehension of different processes allows for the prediction of observable quantities with well-understood model dependencies, as was done in the first prediction the ^{6}He β-decay spectrum to account for two-body current effects.

Over the course of studying the role of two-body currents, the calculations in this thesis highlighted a few important points; namely, (i) that there is a lack of convergence in the power counting of various electroweak observables, (ii) there is an interplay between the $3N$ force and two-body weak currents, and (iii) that this can, in some cases, result in a model sensitivity appearing in the calculation of observable quantities. In the future, detailed studies of weak and electromagnetic observables in the two-body sector should be carried out using a range of different cutoffs in order to illuminate the nature of the model dependencies. Once these issues are studied in further detail, one should revisit the calculations of this thesis to see if the lack of convergence is indeed resolved and/or better understood.

There are clear extensions of thesis that would involve further calculations using QMC with the NV2+3 models. In the moderate momentum transfer regime where the model has now been validated, it would be interesting to perform a QMC evaluation of

G. B. King, *Electroweak Structure and Reactions in Light Nuclei with Quantum Monte Carlo Methods*, Springer Theses, https://doi.org/10.1007/978-3-032-03191-4_7

$0\nu\beta\beta$ matrix elements. While no light systems accessible to the method are candidates for this rare decay, it will be possible to gain insight into how the nuclear dynamics impact the matrix elements by computing the quantity with different model classes for various transitions. It would also be interesting to extend this work to large energy and momentum transfer by validating against quasi-elastic electron scattering data. A recently new QMC technique– the short-time approximation– was developed to address scattering with quasi-elastic kinematics [1, 2]. It will be useful to see how well χEFT models fare at describing this process. Further, it will be a next step to ultimately computing neutrino scattering cross sections relevant to on-going long-baseline searches.

Another future direction of the QMC approach is to push for heavier masses and extend the computations of β-decay, muon capture, and magnetic structure as performed in this thesis to $A \geq 11$ nuclei. In these heavier systems, one will encounter more exotic nuclei, like the ^{11}Li and ^{11}Be ground states. To compute electroweak properties of these halo systems, it may be prudent to use a different approach that accounts for the effect of the continuum. Combining this work with the advances made in computing QMC scattering states [3, 4], one could hope to provide an adequate computation of electroweak properties. Further more, using different model classes to study these states may not only be a way to assess a model dependence, but also be a means to rule out certain model classes if they cannot reproduced important features of these systems like, for example, the parity inversion of the ^{11}Be ground state.

Finally, in light systems, precision β-decay studies have the potential to resolve tension between the SM and measurement or to expose the presence of new physics [5, 6]. In particular, radiative corrections to superallowed $0^+ \to 0^+$ β-decays have received renewed interest [7–9] and could help to understand the present 3σ tension in the top row unitarity of the measured CKM matrix. Given the novel work of computing radiative corrections in a dispersive formalism [7–9] and the possibility of extending χEFT approaches used for neutron β-decay radiative corrections to study nuclear decays [10], there should be QMC calculations of these quantities in both frameworks using the eight NV2+3 model classes. This will provide an assessment of the model dependence of this quantity while also giving a cross-validation of the two approaches.

The studies carried out in this thesis have contributed to the collective efforts of the nuclear theory community to improve our models so that we can maximize the potential of ongoing and planned experimental programs. The future studies proposed in this section will only serve to further bolster the understanding of nuclear theory and make reliable determinations of fundamental physics possible. Given the accuracy of the techniques and the capacity to understand the underlying dynamics, QMC calculations in light nuclei will continue to play a prominent role in these endeavors over the coming years. Therefore, the work carried out in this thesis will serve to guide to guide future QMC and many-body theory contributions in this area and will hopefully inspire improvements to the χEFT approach of modeling nuclear structure and electroweak reactions.

References

1. Saori Pastore, Joseph Carlson, Stefano Gandolfi, Rocco Schiavilla, and Robert B. Wiringa. Quasielastic lepton scattering and back-to-back nucleons in the short-time approximation. 2019
2. Lorenzo Andreoli, Joseph Carlson, Alessandro Lovato, Saori Pastore, Noemi Rocco, and R. B. Wiringa. Electron scattering on A=3 nuclei from quantum Monte Carlo based approaches. *Phys. Rev. C*, 105(1):014002, 2022. 10.1103/PhysRevC.105.014002
3. A.R. Flores, K.M. Nollett, Variational Monte Carlo calculations of n+H3 scattering. Phys. Rev. C **108**(3), 034001 (2023). https://doi.org/10.1103/PhysRevC.108.034001
4. Abraham R Flores. *Quantum Monte Carlo Calculations of Scattering*. PhD thesis, UC Irvine, 2023
5. Vincenzo Cirigliano, Alejandro Garcia, Doron Gazit, Oscar Naviliat-Cuncic, Guy Savard, and Albert Young. Precision Beta Decay as a Probe of New Physics. 7 2019
6. M. Brodeur et al. Nuclear β decay as a probe for physics beyond the Standard Model. 1 2023
7. Chien Yeah Seng, Mikhail Gorchtein, and Michael J. Ramsey-Musolf. Dispersive evaluation of the inner radiative correction in neutron and nuclear β decay. *Phys. Rev. D*, 100(1): 013001, 2019. 10.1103/PhysRevD.100.013001
8. C.-Y. Seng, M. Gorchtein, H.H. Patel, M.J. Ramsey-Musolf, Reduced Hadronic Uncertainty in the Determination of V_{ud}. Phys. Rev. Lett. **121**(24), 241804 (2018). https://doi.org/10.1103/PhysRevLett.121.241804
9. C.-Y. Seng, Radiative Corrections to Semileptonic Beta Decays: Progress and Challenges. Particles **4**(4), 397–467 (2021). https://doi.org/10.3390/particles4040034
10. V. Cirigliano, W. Dekens, E. Mereghetti, O. Tomalak, Effective field theory for radiative corrections to charged-current processes: Vector coupling. Phys. Rev. D **108**(5), 053003 (2023). https://doi.org/10.1103/PhysRevD.108.053003

Appendix A
Expressions of Potential and Current Operators

In this appendix, I detaill the coordinate space expressions of the operators composing the NV2+3 potential. I also provide expressions the vector current operator, as well as the axial charge and current operators. The vector charge operator did not contribute to the observables studied in this thesis and I do not report on it here; however Ref. [1] reports on the momentum space expression of this operator.

A.1 NV2+3 Potential

The expressions of the NV2+3 potential in coordinate space were first reported in Ref. [2]. Here, I list the expressions of the radial functions that are relevant to the studies carried out in this work. First, I begin by describing the radial functions associated with the *NN* interaction that depend on the spacing $r = |\mathbf{r}_i - \mathbf{r}_j|$ between nucleons i and j. The subscripts c, σ, τ, t, and T indicate that the function corresponds to a term involving the operators 1, $\boldsymbol{\sigma}_i \cdot \boldsymbol{\sigma}_j$, $\boldsymbol{\tau}_i \cdot \boldsymbol{\tau}_j$, $S_{ij}(\hat{\mathbf{r}})$, and T_{ij}, respectively. Each term has a superscript denoting what power of the momentum scale Q the operator carries. The OPE contribution to the potential at LO (Q^0) is given by,

$$v^{(0)}_{\sigma T}(r) = \frac{Y_0(r) + 2\,Y_+(r)}{3}\,, \tag{A.1}$$

$$v^{(0)}_{tT}(r) = \frac{T_0(r) + 2\,T_+(r)}{3}\,, \tag{A.2}$$

where,

$$Y_\alpha(r) = \frac{\tilde{g}_A^2}{12\,\pi}\,\frac{m^3_{\pi_\alpha}}{F_\pi^2}\,\frac{e^{-x_\alpha}}{x_\alpha}\,, \tag{A.3}$$

$$T_\alpha(r) = Y_\alpha(r)\left(1 + \frac{3}{x_\alpha} + \frac{3}{x_\alpha^2}\right)\,, \tag{A.4}$$

G. B. King, *Electroweak Structure and Reactions in Light Nuclei with Quantum Monte Carlo Methods*, Springer Theses, https://doi.org/10.1007/978-3-032-03191-4

Table A.1 Values of the charged and neutral pion masses, the proton and neutron masses, the Δ-nucleon mass difference, and electron mass (all in MeV), and of the (adimensional) fine structure constant α used in the NV2+3 potential. Note that $\hbar c$ is taken as 197.32697 MeV fm

m_{π_0}	$m_{\pi_\pm}$	m_n	m_p	$m_{\Delta N}$	m_e	α^{-1}
134.9766	139.5702	939.56524	938.27192	293.1	0.510999	137.03599

Table A.2 Values of "fixed" LECs used in the NV2+3 models: $\tilde{g}_A$, h_A, and c_6 are adimensional, $F_\pi = 2f_\pi$ is given in MeV, and the remaining LECs are given in GeV^{-1}. Note that the value of the nucleon axial coupling $\tilde{g}_A$ used to construct the *NN* potential accounts for the Goldberger-Treiman discrepancy [3, 4]

$\tilde{g}_A$	h_A	F_π	c_1	c_2	c_3	c_4	c_6	$b_3 + b_8$
1.29	2.74	184.80	−0.57	−0.25	−0.79	1.33	5.83	1.40

and $x_\alpha = m_{\pi_\alpha} r$ and F_π is the pion decay amplitude. In Table A.1, I report the values of m_{π_α} and other constants appearing in the NV2+3. Table A.2 reports on F_π and other "fixed" low-energy constants. The charge-dependent radial functions are given by,

$$v_{\rm L}^{\sigma T}(r) = \frac{Y_0(r) - Y_+(r)}{3}\,, \tag{A.5}$$

$$v_{\rm L}^{tT}(r) = \frac{T_0(r) - T_+(r)}{3}\,. \tag{A.6}$$

At next N2LO (Q^2) the long range terms involving no intermedidate Δ-isobars read [5]

$$v_\tau^{(2)}(r;\not\Delta) = \frac{1}{8\pi^3 r^4}\frac{m_\pi}{F_\pi^4}\Big[x\left[1 + 10\tilde{g}_A^2 - \tilde{g}_A^4(23 + 4x^2)\right]K_0(2x) + \left[1 + 2\tilde{g}_A^2(5 + 2x^2) - \tilde{g}_A^4(23 + 12x^2)\right]K_1(2x)\Big]\,, \tag{A.7}$$

$$v_\sigma^{(2)}(r;\not\Delta) = \frac{1}{2\pi^3 r^4}\frac{\tilde{g}_A^4}{F_\pi^4}m_\pi\Big[3x\,K_0(2x) + (3 + 2x^2)K_1(2x)\Big]\,, \tag{A.8}$$

$$v_t^{(2)}(r;\not\Delta) = -\frac{1}{8\pi^3 r^4}\frac{\tilde{g}_A^4}{F_\pi^4}m_\pi\Big[12x\,K_0(2x) + (15 + 4x^2)K_1(2x)\Big]\,, \tag{A.9}$$

where $x = m_\pi r$ (m_π is the average pion mass) and the functions K_n are modified Bessel functions of the second kind, h_A is the N-to-Δ axial coupling, and F_π is the pion decay amplitude. N2LO terms involving one intermediate Δ-isobar are given by

$$v_c^{(2)}(r;\Delta) = -\frac{1}{6\pi^2 r^5\, y}\frac{\tilde{g}_A^2 h_A^2}{F_\pi^4} e^{-2x}\left(6+12x+10x^2+4x^3+x^4\right), \tag{A.10}$$

$$\begin{aligned}
v_\tau^{(2)}(r;\Delta) = &-\frac{1}{216\pi^3 r^5}\frac{h_A^2}{F_\pi^4}\Bigg[\int_0^\infty d\mu \frac{\mu^2}{\sqrt{\mu^2+4x^2}} e^{-\sqrt{\mu^2+4x^2}}(12x^2+5\mu^2+12y^2) \\
&-12y\int_0^\infty d\mu \frac{\mu}{\sqrt{\mu^2+4x^2}} e^{-\sqrt{\mu^2+4x^2}}(2x^2+\mu^2+2y^2)\arctan\frac{\mu}{2y}\Bigg] \\
&-\frac{1}{216\pi^3 r^5}\frac{\tilde{g}_A^2 h_A^2}{F_\pi^4}\Bigg[-\int_0^\infty d\mu \frac{\mu^2}{\sqrt{\mu^2+4x^2}} e^{-\sqrt{\mu^2+4x^2}}(24x^2+11\mu^2+12y^2) \\
&+\frac{6}{y}\int_0^\infty d\mu \frac{\mu}{\sqrt{\mu^2+4x^2}} e^{-\sqrt{\mu^2+4x^2}}(2x^2+\mu^2+2y^2)^2\arctan\frac{\mu}{2y}\Bigg],
\end{aligned} \tag{A.11}$$

$$\begin{aligned}
v_\sigma^{(2)}(r;\Delta) = &-\frac{1}{72\pi^3 r^5}\frac{\tilde{g}_A^2 h_A^2}{F_\pi^4}\Bigg[2\int_0^\infty d\mu \frac{\mu^2}{\sqrt{\mu^2+4x^2}} e^{-\sqrt{\mu^2+4x^2}}(\mu^2+4x^2) \\
&-\frac{1}{y}\int_0^\infty d\mu \frac{\mu}{\sqrt{\mu^2+4x^2}} e^{-\sqrt{\mu^2+4x^2}}(\mu^2+4x^2)(\mu^2+4y^2)\arctan\frac{\mu}{2y}\Bigg],
\end{aligned} \tag{A.12}$$

$$v_{\sigma\tau}^{(2)}(r;\Delta) = \frac{1}{54\pi^2 r^5\, y}\frac{\tilde{g}_A^2 h_A^2}{F_\pi^4} e^{-2x}(1+x)\left(3+3x+x^2\right), \tag{A.13}$$

$$\begin{aligned}
v_t^{(2)}(r;\Delta) = &\frac{1}{144\pi^3 r^5}\frac{\tilde{g}_A^2 h_A^2}{F_\pi^4}\Bigg[2\int_0^\infty d\mu \frac{\mu^2}{\sqrt{\mu^2+4x^2}} e^{-\sqrt{\mu^2+4x^2}}(3+3\sqrt{\mu^2+4x^2}+\mu^2+4x^2) \\
&-\frac{1}{y}\int_0^\infty d\mu \frac{\mu}{\sqrt{\mu^2+4x^2}} e^{-\sqrt{\mu^2+4x^2}}(\mu^2+4y^2) \\
&\times(3+3\sqrt{\mu^2+4x^2}+\mu^2+4x^2)\arctan\frac{\mu}{2y}\Bigg],
\end{aligned} \tag{A.14}$$

$$v_{t\tau}^{(2)}(r;\Delta) = -\frac{1}{54\pi^2 r^5\, y}\frac{\tilde{g}_A^2 h_A^2}{F_\pi^4} e^{-2x}(1+x)\left(3+3x+2x^2\right), \tag{A.15}$$

where $y = m_{N\Delta} r$ with $m_{N\Delta}$ being the Δ-nucleon mass difference. The integrations over μ are carried out numerically to obtain the expressions used in QMC codes. Finally, the long-range terms involving two intermediate Δ-isobars are given by,

$$\begin{aligned}
v_c^{(2)}(r;2\Delta) = &-\frac{1}{108\pi^3 r^5}\frac{h_A^4}{F_\pi^4}\Bigg[\int_0^\infty d\mu \frac{\mu^2}{\sqrt{\mu^2+4x^2}} e^{-\sqrt{\mu^2+4x^2}}\left[4y^2+2\frac{(2x^2+\mu^2+2y^2)^2}{(\mu^2+4y^2)}\right] \\
&+\frac{1}{y}\int_0^\infty d\mu \frac{\mu}{\sqrt{\mu^2+4x^2}} e^{-\sqrt{\mu^2+4x^2}} \\
&(2x^2+\mu^2+2y^2)(2x^2+\mu^2-6y^2)\arctan\frac{\mu}{2y}\Bigg],
\end{aligned} \tag{A.16}$$

$$\begin{aligned} v_T^{(2)}(r;2\Delta) = &-\frac{1}{1944\pi^3 r^5}\frac{h_A^4}{F_\pi^4}\Bigg[\int_0^\infty d\mu \frac{\mu^2}{\sqrt{\mu^2+4x^2}}e^{-\sqrt{\mu^2+4x^2}} \\ &\times\left[(24x^2+11\mu^2+24y^2)+6\frac{(2x^2+\mu^2+2y^2)^2}{(\mu^2+4y^2)}\right] \\ &-\frac{3}{y}\int_0^\infty d\mu \frac{\mu}{\sqrt{\mu^2+4x^2}}e^{-\sqrt{\mu^2+4x^2}} \\ &\times(2x^2+\mu^2+2y^2)(2x^2+\mu^2+10y^2)\arctan\frac{\mu}{2y}\Bigg], \end{aligned} \tag{A.17}$$

$$\begin{aligned} v_\sigma^{(2)}(r;2\Delta) = &-\frac{1}{1296\pi^3 r^5}\frac{h_A^4}{F_\pi^4}\Bigg[-6\int_0^\infty d\mu \frac{\mu^2}{\sqrt{\mu^2+4x^2}}e^{-\sqrt{\mu^2+4x^2}}(\mu^2+4x^2) \\ &+\frac{1}{y}\int_0^\infty d\mu \frac{\mu}{\sqrt{\mu^2+4x^2}}e^{-\sqrt{\mu^2+4x^2}}(\mu^2+4x^2)(\mu^2+12y^2)\arctan\frac{\mu}{2y}\Bigg], \end{aligned} \tag{A.18}$$

$$\begin{aligned} v_{\sigma\tau}^{(2)}(r;2\Delta) = &-\frac{1}{7776\pi^3 r^5}\frac{h_A^4}{F_\pi^4}\Bigg[-2\int_0^\infty d\mu \frac{\mu^2}{\sqrt{\mu^2+4x^2}}e^{-\sqrt{\mu^2+4x^2}}(\mu^2+4x^2) \\ &+\frac{1}{y}\int_0^\infty d\mu \frac{\mu}{\sqrt{\mu^2+4x^2}}e^{-\sqrt{\mu^2+4x^2}}(\mu^2+4x^2)(-\mu^2+4y^2)\arctan\frac{\mu}{2y}\Bigg], \end{aligned} \tag{A.19}$$

$$\begin{aligned} v_t^{(2)}(r;2\Delta) = &\frac{1}{2592\pi^3 r^5}\frac{h_A^4}{F_\pi^4}\Bigg[-6\int_0^\infty d\mu \frac{\mu^2}{\sqrt{\mu^2+4x^2}}e^{-\sqrt{\mu^2+4x^2}}(3+3\sqrt{\mu^2+4x^2}+\mu^2+4x^2) \\ &+\frac{1}{y}\int_0^\infty d\mu \frac{\mu}{\sqrt{\mu^2+4x^2}}e^{-\sqrt{\mu^2+4x^2}} \\ &\times(3+3\sqrt{\mu^2+4x^2}+\mu^2+4x^2)(\mu^2+12y^2)\arctan\frac{\mu}{2y}\Bigg], \end{aligned} \tag{A.20}$$

$$\begin{aligned} v_{t\tau}^{(2)}(r;2\Delta) = &\frac{1}{15552\pi^3 r^5}\frac{h_A^4}{F_\pi^4}\Bigg[-2\int_0^\infty d\mu \frac{\mu^2}{\sqrt{\mu^2+4x^2}}e^{-\sqrt{\mu^2+4x^2}}(3+3\sqrt{\mu^2+4x^2}+\mu^2+4x^2) \\ &+\frac{1}{y}\int_0^\infty d\mu \frac{\mu}{\sqrt{\mu^2+4x^2}}e^{-\sqrt{\mu^2+4x^2}}(3+3\sqrt{\mu^2+4x^2}+\mu^2+4x^2) \\ &\times(-\mu^2+4y^2)\arctan\frac{\mu}{2y}\Bigg]. \end{aligned} \tag{A.21}$$

In addition to the long-range operators, there are several short-range contact terms containing the LECs which must be fit to data. The radial dependence of the contact operators arises from the substitution,

$$\delta(r) \to C_{R_s}(r) = \frac{1}{\pi^{3/2}R_S^3}e^{-(r/R_S)^2} \tag{A.22}$$

which, turn, result in dependencies on the derivatives of Gaussians,

$$C_{R_S}^{(n)} = \frac{d^n C_{R_s}(r)}{dr^n}, \tag{A.23}$$

when the Fourier transform is performed [2]. In addition to the dependencies on the pion exchange operator structures, there are additional subscripts $b,\ bb,\ q,$ and τz

indicating that the radial functions are associated with the operators $\mathbf{L}\cdot\mathbf{S}$, $(\mathbf{L}\cdot\mathbf{S})^2$, $\mathbf{L}^2$, and $\tau_{i,z}+\tau_{j,z}$, respectively. The LECs C_S and C_T enter at leading order, the C_i's at N2LO, and the D_i's at N3LO in the charge-independent potential. Charge-dependent LECs $C_i^{\rm IT}$ and $C_i^{\rm IV}$ come from symmetry breaking at LO and N2LO. The charge-dependent terms that would arise from the N3LO potential are neglected. The radial potentials in terms of these LECs are,

$$v_S^c(r) = C_S\, C_{R_S}(r) + C_1\left[-C_{R_S}^{(2)}(r) - \frac{2}{r}\, C_{R_S}^{(1)}(r)\right] + D_1\left[C_{R_S}^{(4)}(r) + \frac{4}{r}\, C_{R_S}^{(3)}(r)\right], \quad (A.24)$$

$$v_S^\tau(r) = C_2\left[-C_{R_S}^{(2)}(r) - \frac{2}{r}\, C_{R_S}^{(1)}(r)\right] + D_2\left[C_{R_S}^{(4)}(r) + \frac{4}{r}\, C_{R_S}^{(3)}(r)\right], \quad (A.25)$$

$$v_S^\sigma(r) = C_T\, C_{R_S}(r) + C_3\left[-C_{R_S}^{(2)}(r) - \frac{2}{r}\, C_{R_S}^{(1)}(r)\right] + D_3\left[C_{R_S}^{(4)}(r) + \frac{4}{r}\, C_{R_S}^{(3)}(r)\right], \quad (A.26)$$

$$v_S^{\sigma\tau}(r) = C_4\left[-C_{R_S}^{(2)}(r) - \frac{2}{r}\, C_{R_S}^{(1)}(r)\right] + D_4\left[C_{R_S}^{(4)}(r) + \frac{4}{r}\, C_{R_S}^{(3)}(r)\right], \quad (A.27)$$

$$\begin{aligned} v_S^t(r) &= -C_5\left[C_{R_S}^{(2)}(r) - \frac{1}{r}\, C_{R_S}^{(1)}(r)\right] \\ &\quad + D_5\left[C_{R_S}^{(4)}(r) + \frac{1}{r} C_{R_S}^{(3)}(r) - \frac{6}{r^2} C_{R_S}^{(2)}(r) + \frac{6}{r^3} C_{R_S}^{(1)}(r)\right], \end{aligned} \quad (A.28)$$

$$\begin{aligned} v_S^{t\tau}(r) &= -C_6\left[C_{R_S}^{(2)}(r) - \frac{1}{r}\, C_{R_S}^{(1)}(r)\right] \\ &\quad + D_6\left[C_{R_S}^{(4)}(r) + \frac{1}{r} C_{R_S}^{(3)}(r) - \frac{6}{r^2} C_{R_S}^{(2)}(r) + \frac{6}{r^3} C_{R_S}^{(1)}(r)\right], \end{aligned} \quad (A.29)$$

$$v_S^b(r) = -C_7 \frac{1}{r} C_{R_S}^{(1)}(r) + D_7\left[\frac{1}{r} C_{R_S}^{(3)}(r) + 2\,\frac{1}{r^2} C_{R_S}^{(2)}(r) - \frac{2}{r^3} C_{R_S}^{(1)}(r)\right], \quad (A.30)$$

$$v_S^{b\tau}(r) = D_8\left[\frac{1}{r} C_{R_S}^{(3)}(r) + 2\,\frac{1}{r^2} C_{R_S}^{(2)}(r) - \frac{2}{r^3} C_{R_S}^{(1)}(r)\right], \quad (A.31)$$

$$v_S^{bb}(r) = -D_9 \frac{1}{r^2}\left[C_{R_S}^{(2)}(r) - \frac{1}{r}\, C_{R_S}^{(1)}(r)\right], \quad (A.32)$$

$$v_S^q(r) = -D_{10} \frac{1}{r^2}\left[C_{R_S}^{(2)}(r) - \frac{1}{r}\, C_{R_S}^{(1)}(r)\right], \quad (A.33)$$

$$v_S^{q\sigma}(r) = -D_{11} \frac{1}{r^2}\left[C_{R_S}^{(2)}(r) - \frac{1}{r}\, C_{R_S}^{(1)}(r)\right], \quad (A.34)$$

$$v_S^T(r) = C_0^{\rm IT}\, C_{R_S}(r) + C_1^{\rm IT}\left[-C_{R_S}^{(2)}(r) - \frac{2}{r}\, C_{R_S}^{(1)}(r)\right], \quad (A.35)$$

$$v_S^{\tau z}(r) = C_0^{\rm IV}\, C_{R_S}(r) \quad (A.36)$$

$$v_S^{\sigma T}(r) = C_2^{\rm IT}\left[-C_{R_S}^{(2)}(r) - \frac{2}{r}\, C_{R_S}^{(1)}(r)\right], \quad (A.37)$$

$$v_S^{tT}(r) = -C_3^{\rm IT}\left[C_{R_S}^{(2)}(r) - \frac{1}{r}\, C_{R_S}^{(1)}(r)\right], \quad (A.38)$$

$$v_S^{bT}(r) = -C_4^{\rm IT} \frac{1}{r} C_{R_S}^{(1)}(r)\,. \quad (A.39)$$

Table A.3 Values of the LECs corresponding to potential models Ia and Ib (fitted up to 125 MeV lab energy). The notation $(\pm n)$ indicates $\times 10^{\pm n}$. Values originally reported in Ref. [6]

LECs	Model Ia	Model Ib
C_S (fm^2)	0.2726141(+1)	0.8038124(+1)
C_T (fm^2)	–0.5228448	–0.1203741(+1)
C_1 (fm^4)	–0.6992838(–1)	–0.2280422
C_2 (fm^4)	–0.1496013	–0.2249889
C_3 (fm^4)	–0.2502401(–1)	–0.4007665(–1)
C_4 (fm^4)	–0.2728396(–1)	0.1243960(–1)
C_5 (fm^4)	–0.6530008(–2)	–0.1870727(–1)
C_6 (fm^4)	–0.7554924(–1)	–0.7406609(–1)
C_7 (fm^4)	–0.1017206(+1)	–0.1197452(+1)
D_1 (fm^6)	–0.4251199(–1)	–0.3820959(–1)
D_2 (fm^6)	–0.5567938(–2)	–0.5343034(–2)
D_3 (fm^6)	–0.1666607(–1)	–0.1601394(–1)
D_4 (fm^6)	0.1054347(–2)	0.4219347(–2)
D_5 (fm^6)	0.5383828(–2)	0.8971752(–2)
D_6 (fm^6)	–0.8012050(–2)	–0.5986245(–2)
D_7 (fm^6)	–0.2309392(–1)	–0.6180197(–2)
D_8 (fm^6)	0.1383136(–1)	0.1782567(–1)
D_9 (fm^6)	0.4797012(–1)	0.3094851(–1)
D_{10} (fm^6)	–0.1156876	–0.8073891(–1)
D_{11} (fm^6)	–0.1453295(–1)	–0.1162060(–1)
C_0^{IV} (fm^2)	0.9325477(–2)	0.1018989(–1)
C_0^{IT} (fm^2)	0.1578240(–1)	0.2416591(–1)
C_1^{IT} (fm^4)	–0.2179452(–2)	–0.3707396(–2)
C_2^{IT} (fm^4)	–0.6288540(–2)	–0.3601899(–2)
C_3^{IT} (fm^4)	–0.5799803(–2)	–0.4559006(–2)
C_4^{IT} (fm^4)	0.2250167(–1)	0.1859997(–1)

The 26 LECs in this expression are unknown and were fit to data in Ref. [6]. Tables A.3 and A.4 report on the values fitted for model classes I and II, respectively.

The coordinate space expression of the electromagnetic potential is the same as for the Argonne interactions [7]. In addition to radial and parametric dependencies, the potential also depends on the particular *NN* channel that is interacting. In the *pp* system, the potential includes one–(C1) and two-photon Coulomb (C2) terms, the Darwin-Foldy (DF) term, vacuum polarization (VP), and the magnetic moment (MM) interaction,

$$v^{\mathrm{EM}}(pp) = V_{C1}(pp) + V_{C2} + V_{\mathrm{DF}} + V_{\mathrm{VP}} + V_{\mathrm{MM}}(pp)\,, \tag{A.40}$$

Table A.4 Values of the LECs corresponding to potential models IIa and IIb (fitted up to 200 MeV lab energy). The notation $(\pm n)$ indicates $\times 10^{\pm n}$. Values originally reported in Ref. [6]

LECs	Model IIa	Model IIb
C_S (fm^2)	0.2936041(+1)	0.8398499(+1)
C_T (fm^2)	−0.4933897	−0.1207696(+1)
C_1 (fm^4)	−0.1013462	−0.2324413
C_2 (fm^4)	−0.1444844	−0.2108143
C_3 (fm^4)	−0.3647634(−1)	−0.3461629(−1)
C_4 (fm^4)	−0.1630825(−1)	0.8748772(−2)
C_5 (fm^4)	−0.6658100(−2)	−0.3614304(−1)
C_6 (fm^4)	−0.6176835(−1)	−0.5542581(−1)
C_7 (fm^4)	−0.9578191	−0.1019849(+1)
D_1 (fm^6)	−0.3102824(−1)	−0.1193597(−1)
D_2 (fm^6)	−0.4438695(−2)	−0.4450346(−2)
D_3 (fm^6)	−0.1351171(−1)	−0.9542801(−2)
D_4 (fm^6)	−0.7084459(−3)	0.3976205(−2)
D_5 (fm^6)	0.1110108(−1)	0.7809205(−2)
D_6 (fm^6)	−0.8598857(−2)	−0.7362895(−2)
D_7 (fm^6)	−0.5367908(−1)	−0.4158494(−2)
D_8 (fm^6)	0.3119241(−1)	0.1090986(−1)
D_9 (fm^6)	0.3281636(−1)	0.6095858(−3)
D_{10} (fm^6)	−0.8647128(−1)	−0.5432144(−1)
D_{11} (fm^6)	−0.1167788(−1)	−0.5186422(−2)
$C_0^{\rm IV}$ (fm^2)	0.9575695(−2)	0.1077541(−1)
$C_0^{\rm IT}$ (fm^2)	0.2194758(−1)	0.2102140(−1)
$C_1^{\rm IT}$ (fm^4)	−0.1550501(−2)	0.1152693(−3)
$C_2^{\rm IT}$ (fm^4)	−0.8354679(−2)	−0.1391786(−2)
$C_3^{\rm IT}$ (fm^4)	−0.6682746(−2)	−0.3194459(−3)
$C_4^{\rm IT}$ (fm^4)	0.1276971(−1)	0.2879873(−2)

with the potentials defined as,

$$V_{C1}(pp) = \alpha \frac{F_C(r)}{r} , \tag{A.41}$$

$$V_{C2} = -\frac{\alpha}{2m_p^2}\left[(\nabla^2+k^2)\frac{F_C(r)}{r} + \frac{F_C(r)}{r}(\nabla^2+k^2)\right] \approx -\frac{\alpha^2}{m_p}\left[\frac{F_C(r)}{r}\right]^2 , \tag{A.42}$$

$$V_{\rm DF} = -\frac{\alpha}{4m_p^2} F_\delta(r) , \tag{A.43}$$

$$V_{\rm VP} = \frac{2\alpha^2}{3\pi}\frac{F_C(r)}{r}\int_1^\infty d\mu\, e^{-2m_e r\mu}\left[1+\frac{1}{2\mu^2}\right]\frac{\sqrt{\mu^2-1}}{\mu^2} , \tag{A.44}$$

$$V_{\rm MM}(pp) = -\frac{\alpha}{4m_p^2}\mu_p^2\left[\frac{2}{3}F_\delta(r)\boldsymbol{\sigma}_i\cdot\boldsymbol{\sigma}_j + \frac{F_t(r)}{r^3}S_{ij}\right] - \frac{\alpha}{2m_p^2}(4\mu_p - 1)\frac{F_{ls}(r)}{r^3}\mathbf{L}\cdot\mathbf{S}\,, \tag{A.45}$$

and α being the fine structure constant, m_p the proton mass, $\mu_p = 2.7929$ n. m. is the proton magnetic moment, and m_e is the electron mass. The form factors are defined (for $b = 4.27\ \text{fm}^{-1}$),

$$\begin{aligned}
F_C(r) &= 1 - \left(1 + \frac{11}{16}x + \frac{3}{16}x^2 + \frac{1}{48}x^3\right)e^{-x}\,,\\
F_\delta(r) &= b^3\left(\frac{1}{16} + \frac{1}{16}x + \frac{1}{48}x^2\right)e^{-x}\,,\\
F_t(r) &= 1 - \left(1 + x + \frac{1}{2}x^2 + \frac{1}{6}x^3 + \frac{1}{24}x^4 + \frac{1}{144}x^5\right)e^{-x}\,,\\
F_{ls}(r) &= 1 - \left(1 + x + \frac{1}{2}x^2 + \frac{7}{48}x^3 + \frac{1}{48}x^4\right)e^{-x}\,.
\end{aligned} \tag{A.46}$$

with $x = br$. In the np system, there is only the C1 term arising from the the interaction of the proton with the neutron charge distribution as well as the MM interaction,

$$v^{EM}(np) = V_{C1}(np) + V_{\rm MM}(np)\,. \tag{A.47}$$

The expressions are now,

$$V_{C1}(np) = \alpha\beta_n\frac{F_{np}(r)}{r}\,, \tag{A.48}$$

$$V_{\rm MM}(np) = -\frac{\alpha}{4m_n m_p}\mu_n\mu_p\left[\frac{2}{3}F_\delta(r)\boldsymbol{\sigma}_i\cdot\boldsymbol{\sigma}_j + \frac{F_t(r)}{r^3}S_{ij}\right] - \frac{\alpha}{2m_n m_r}\mu_n\frac{F_{ls}(r)}{r^3}(\mathbf{L}\cdot\mathbf{S} + \mathbf{L}\cdot\mathbf{A})\,, \tag{A.49}$$

where $\beta_n = 0.0189\ \text{fm}^2$ was extracted from neutron-noble gas scattering assuming a modified Galster form factor for the neutron charge distribution [8], $\mu_n = -1.9131$ n. m. is the neutron magnetic moment, m_r is the reduced mass of the np system, and $\mathbf{A} = (\boldsymbol{\sigma}_i - \boldsymbol{\sigma}_j)/2$. Finally, in the nn system, the Coulomb term from the interaction of the neutron charge distributions is dropped, leaving only the MM interaction,

$$\begin{aligned}
v^{EM}(nn) &= V_{MM}(nn)\\
&= -\frac{\alpha}{4m_n^2}\mu_n^2\left[\frac{2}{3}F_\delta(r)\boldsymbol{\sigma}_i\cdot\boldsymbol{\sigma}_j + \frac{F_t(r)}{r^3}S_{ij}\right]\,.
\end{aligned} \tag{A.50}$$

In addition to the two-body potential, one includes a three-body force, as well. Ref. [9] first reported the form of the potential used in the Norfolk model. The potential may be written as the sum of a long-range TPE term and short-range CT term,

$$V_{ijk} = \sum_{\text{cyclic } ijk} \left(V_{ijk}^{2\pi} + V_{ijk}^{\text{CT}} \right) . \tag{A.51}$$

The two-pion range term includes the exchange of an intermediate Δ plus a term from interactions proportional to the LECs c_1, c_3, and c_4 in the sub-leading chiral Lagrangian [10],

$$V_{ijk}^{2\pi} = \frac{\widetilde{g}_A^2}{256\,\pi^2} \frac{m_\pi^6}{f_\pi^4} \Big[8\,c_1\, \Sigma_{ij}\, \Sigma_{kj}\, \mathcal{T}_{ijk}^{(+)} + \frac{2}{9} \widetilde{c}_3\, \Sigma_{ijk}^{(+)}\, \mathcal{T}_{ijk}^{(+)}$$

$$-\frac{1}{9}\Big(\widetilde{c}_4 + \frac{1}{4m}\Big) \Sigma_{ijk}^{(-)}\, \mathcal{T}_{ijk}^{(-)} \Big] , \tag{A.52}$$

with spin and isospin operator structures defined, respectively, as $\Sigma_{lm} \equiv \widetilde{Z}_\pi(r_{lm})\, \boldsymbol{\sigma}_l \cdot \hat{\mathbf{r}}_{lm}$, where $\mathbf{r}_{lm} \equiv \mathbf{r}_l - \mathbf{r}_m$, and

$$\Sigma_{ijk}^{(\mp)} \equiv \big[\widetilde{X}_{ij}\,,\, \widetilde{X}_{jk}\big]_\mp\,, \quad \mathcal{T}_{ijk}^{(\mp)} \equiv \big[\boldsymbol{\tau}_i \cdot \boldsymbol{\tau}_j\,,\, \boldsymbol{\tau}_j \cdot \boldsymbol{\tau}_k\big]_\mp\,, \tag{A.53}$$

$$\widetilde{X}_{ij} \equiv \widetilde{T}_\pi(r_{ij})\, S_{ij} + \widetilde{Y}_\pi(r_{ij})\, \boldsymbol{\sigma}_i \cdot \boldsymbol{\sigma}_j\,, \tag{A.54}$$

$$\widetilde{Z}_\pi(r) \equiv -\left(1 + \frac{1}{m_\pi\, r}\right) \widetilde{Y}_\pi(r)\,, \tag{A.55}$$

The $[\ldots\,,\,\ldots]_\mp$ are commutators $(-)$ or anti-commutators $(+)$, $f_\pi = F_\pi/2$, and the $\widetilde{Y}_\pi$ and $\widetilde{Y}_\pi$ indicate the Yukawa functions in Eqs. A.1 and A.2 regulated by the long-range regulator in Eq. 2.10. The LECs $\widetilde{c}_3$ and $\widetilde{c}_4$ are related to c_3 and c_4 via,

$$\widetilde{c}_3 = c_3 - \frac{h_A^2}{9\, m_{\Delta N}} \tag{A.56}$$

$$\widetilde{c}_4 = c_4 + \frac{h_A^2}{18\, m_{\Delta N}} . \tag{A.57}$$

The CT term is given by,

$$V_{ijk}^{\text{CT}} = \frac{\widetilde{g}_A\, c_D}{96\,\pi} \frac{m_\pi^3}{\Lambda_\chi\, f_\pi^4} \boldsymbol{\tau}_i \cdot \boldsymbol{\tau}_k\, \widetilde{X}_{ik} \left[C_{R_S}(r_{ij}) + C_{R_S}(r_{jk}) \right]$$

$$+ \frac{c_E}{\Lambda_\chi\, f_\pi^4} \boldsymbol{\tau}_i \cdot \boldsymbol{\tau}_k\, C_{R_S}(r_{ij})\, C_{R_S}(r_{jk})\,, \tag{A.58}$$

where Λ_χ is the chiral-symmetry-breaking scale assumed to be $\Lambda_\chi = 1$ GeV and the unknown LECs c_D and c_E must be fit to data. Their values are listed in Table A.5.

Table A.5 The values of LECs c_D and c_E in the NV3 interactions obtained from fits to either the *nd* scattering length and trinucleon binding energies [9] or to the central value of the ^{3}H GT matrix element and the trinucleon binding energies (starred values) [11]

	Ia (Ia*)	Ib (Ib*)	IIa (IIa*)	IIb (IIb*)
c_D	3.666 (–0.635)	–2.061 (–4.71)	1.278 (–0.61)	–4.480 (–5.25)
c_E	–1.638 (–0.090)	–0.982 (0.55)	–1.029 (–0.35)	–0.412 (0.05)

A.2 Vector Current Operator

The vector current operator from TOPT was first derived in Ref. [12] and the Δ-full current operators consistent with the NV2+3 were most recently reported in Refs. [13] and [14]. The LO current (Q^{-2}) is given by,

$$\mathbf{j}^{(-2)}(\mathbf{q}) = \frac{\epsilon_i(q_\mu^2)}{2m}\left[\mathbf{p}_i\ ,\ e^{i\mathbf{q}\cdot\mathbf{r}_i}\right]_+ + i\,\frac{\mu_i(q_\mu^2)}{2m}\ e^{i\mathbf{q}\cdot\mathbf{r}_i}\ \boldsymbol{\sigma}_i\times\mathbf{q}\ , \tag{A.59}$$

where $\mathbf{p}_i = -i\,\nabla_i$ and m is the average nucleon mass. The operators ϵ_i and μ_i are isospin dependent quantities defined as,

$$\epsilon_i = \frac{1}{2}\left(G_E^S(q_\mu^2) + G_E^V(q_\mu^2)\tau_{i,z}\right)\ , \tag{A.60}$$

$$\mu_i = \frac{1}{2}\left(G_M^S(q_\mu^2) + G_M^V(q_\mu^2)\tau_{i,z}\right)\ . \tag{A.61}$$

Here e is the elementary charge and the isoscalar/isovector combinations of the proton (p) and neutron (n) electric (E) and magnetic (M) form factors are defined as

$$G_{E/M}^{S/V} = G_{E/M}^p(q_\mu^2) \pm G_{E/M}^n(q_\mu^2)\ , \tag{A.62}$$

which depend on $q_\mu = q^2 - \omega^2$, with ω being the energy transfer. Though the form factors, which appear in all order of the the current, themselves have a power series expansion in the low momentum scale Q of χEFT, this is typically neglected and they are treated as multiplicative factors. The form factors in this work are not taken from chiral perturbation theory, but rather from experimental fits of elastic scattering on the proton and deuteron. I adopt the same form factors as in Ref. [14], which are given by

$$\begin{aligned}
G_E^p(q_\mu^2) &= G_D(q_\mu^2)\ ,\\
G_E^n(q_\mu^2) &= -\mu^n\chi G_D(q_\mu^2)(1+4\chi)^{-1}\ ,\\
G_M^p(q_\mu^2) &= \mu_p G_D(q_\mu^2)\ ,\\
G_M^n(q_\mu^2) &= \mu_n G_D(q_\mu^2)\ ,
\end{aligned}$$

where μ_p and μ_n are the proton and neutron magnetic moments, respectively, $\chi \equiv q_\mu^2/(4m_N^2)$, and the dipole form factor is defined as

$$G_D(q_\mu^2) = (1 + q_\mu^2/\Lambda_D^2)^{-2}\,, \tag{A.63}$$

for $\Lambda_D = 0.83$ GeV.

At NLO (Q^{-1}), the currents receive a long-range OPE contributions given by,

$$\mathbf{j}_{\rm OPE}^{\rm NLO}(\mathbf{q}) = G_E^V(q_\mu^2) \times \tag{A.64}$$

$$\Big[\mathrm{e}^{i\,\mathbf{q}\cdot\mathbf{r}_i}\,(\boldsymbol{\tau}_i \times \boldsymbol{\tau}_j)_z\, I_0^\pi(x_{ij})\,\boldsymbol{\sigma}_i\;\boldsymbol{\sigma}_j\cdot\hat{\mathbf{r}}_{ij} + (i \rightleftharpoons j)$$
$$+\,\mathrm{e}^{i\,\mathbf{q}\cdot\mathbf{R}_{ij}}\,(\boldsymbol{\tau}_i \times \boldsymbol{\tau}_j)_z\;\boldsymbol{\sigma}_i\cdot\left(\boldsymbol{\nabla}_{ij}^x + i\,\frac{\mathbf{q}}{2\,m_\pi}\right)$$
$$\times\,\boldsymbol{\sigma}_j\cdot\left(\boldsymbol{\nabla}_{ij}^x - i\,\frac{\mathbf{q}}{2\,m_\pi}\right)\boldsymbol{\nabla}_{ij}^x\, L_0^\pi(\boldsymbol{x}_{ij},\mathbf{q})\Big]\,, \tag{A.65}$$

with $\mathbf{r}_{ij}$ indicating the relative coordinate and $\mathbf{R}_{ij} = (\mathbf{r}_i + \mathbf{r}_j)/2$. The gradients $\boldsymbol{\nabla}_{ij}^\mu$ are taken with respect to the adimensional variables $\boldsymbol{\mu}_{ij} = m_\pi \mathbf{r}_{ij}$, and the correlation functions $I_0^\pi(\mu)$ and $L_0^\pi(\boldsymbol{\mu},\mathbf{q})$ are defined as

$$I_0^\pi(x) = -\frac{\tilde{g}_A^2}{16\pi}\,\frac{m_\pi^2}{f_\pi^2}\,(1+x)\,\frac{e^{-x}}{x^2}\,, \tag{A.66}$$

$$L_0^\pi(\boldsymbol{x},\mathbf{q}) = \frac{\tilde{g}_A^2}{16\pi}\,\frac{m_\pi^2}{f_\pi^2}\int_{-1/2}^{1/2} dz\;\mathrm{e}^{-i\,z(\mathbf{q}/m_\pi)\cdot\boldsymbol{x}}\;\frac{e^{-x\,L(z,q)}}{L(z,q)}\,, \tag{A.67}$$

using the function,

$$L(z,q) = \sqrt{1 + \frac{q^2}{4\,m_\pi^2}\,(1 - 4\,z^2)}\,. \tag{A.68}$$

The N2LO current involves a one-body relativistic corrections (RC) and a term of one-pion range involving the excitation of an intermediate Δ,

$$\mathbf{j}_{\rm RC}^{(0)}(\mathbf{q}) = \frac{\epsilon_i(q_\mu^2)}{16m^3}\left[2\left(p_i^2 + \frac{q^2}{4}\right)(2\,\mathbf{p}_i + i\,\boldsymbol{\sigma}_i\times\mathbf{q}) + \mathbf{p}_i\cdot\mathbf{q}\;(\mathbf{q} + 2\,i\,\boldsymbol{\sigma}_i\times\mathbf{p}_i)\;,\;\mathrm{e}^{i\mathbf{q}\cdot\mathbf{r}_i}\right]_+$$
$$-i\,\frac{\mu_i(q_\mu^2) - \epsilon_i(q_\mu^2)}{16m^3}\,\big[\mathbf{p}_i\cdot\mathbf{q}(4\,\boldsymbol{\sigma}_i\times\mathbf{p}_i - i\,\mathbf{q})$$
$$-\,(2\,i\,\mathbf{p}_i - \boldsymbol{\sigma}_i\times\mathbf{q})\,\frac{q^2}{2} + 2\,(\mathbf{p}_i\times\mathbf{q})\,\boldsymbol{\sigma}_i\cdot\mathbf{p}_i\;,\;\mathrm{e}^{i\mathbf{q}\cdot\mathbf{r}_i}\Big]_+\,, \tag{A.69}$$

$$\mathbf{j}_\Delta^{(0)}(\mathbf{q}) = -iG_{\gamma N\Delta}(q_{mu}^2)\Big\{\mathrm{e}^{i\mathbf{q}\cdot\mathbf{r}_i}\,\tau_{j,z}\left[I_1^\Delta(x_{ij})\,\boldsymbol{\sigma}_j + I_2^\Delta(x_{ij})\,\boldsymbol{\sigma}_j\cdot\hat{\mathbf{r}}_{ij}\,\hat{\mathbf{r}}_{ij}\right]$$

$$-\frac{1}{4}\,\mathrm{e}^{i\,\mathbf{q}\cdot\mathbf{r}_i}\,\left(\boldsymbol{\tau}_i\times\boldsymbol{\tau}_j\right)_z\left[I_1^{\Delta}(x_{ij})\,\boldsymbol{\sigma}_i\times\boldsymbol{\sigma}_j+I_2^{\Delta}(x_{ij})\,\boldsymbol{\sigma}_j\cdot\hat{\mathbf{r}}_{ij}\;\boldsymbol{\sigma}_i\times\hat{\mathbf{r}}_{ij}\right]\Bigg\}\times\frac{\mathbf{q}}{m_\pi}$$
$$+(i\rightleftharpoons j)\,, \tag{A.70}$$

with the correlation functions $I_k^{\Delta}(\mu)$ defined as follows,

$$I_1^{\Delta}(\mu)=-\left(\frac{\tilde{g}_A\,h_A}{18\,\pi}\,\frac{1}{2m}\,\frac{m_\pi^2}{m_{\Delta N}}\,\frac{m_\pi^2}{f_\pi^2}\right)(1+\mu)\,\frac{e^{-\mu}}{\mu^3}\,, \tag{A.71}$$

$$I_2^{\Delta}(\mu)=\left(\frac{\tilde{g}_A\,h_A}{18\,\pi}\,\frac{1}{2m}\,\frac{m_\pi^2}{m_{\Delta N}}\,\frac{m_\pi^2}{f_\pi^2}\right)(3+3\,\mu+\mu^2)\,\frac{e^{-\mu}}{\mu^3}\,, \tag{A.72}$$

where $M_{\Delta N}$ is the N to Δ mass difference and $h_A=2.73$ is the $\pi N\Delta$ coupling constant. The N-to-Δ transition form factor is defined as,

$$G_{\gamma N\Delta}(q_\mu^2)=\frac{\mu_{\gamma N\Delta}}{(1+q_\mu^2/\Lambda_{\Delta,1}^2)\sqrt{1+q_\mu^2/\Lambda_{\Delta,2}^2}}\,, \tag{A.73}$$

where the transition magnetic moment $\mu_{\gamma N\Delta}=3\mu_N$ is taken from an analysis of γN data in the Δ resonance region [15] and I adopt $\Lambda_{\Delta,1}=0.84$ GeV and $\Lambda_{\Delta,2}=1.2$ GeV from the same analysis.

At N3LO (Q^1), the long-range current operator can be broken down into an IS OPE term

$$\mathbf{j}^{(1)}_{\rm IS\;OPE}(\mathbf{q})=-i\,\mathrm{e}^{i\mathbf{q}\cdot\mathbf{r}_i}\,G_{\gamma\pi\rho}(q_\mu^2)\,\boldsymbol{\tau}_i\cdot\boldsymbol{\tau}_j\left[I_1^S(x_{ij})\,\boldsymbol{\sigma}_j+I_2^S(x_{ij})\;\boldsymbol{\sigma}_j\cdot\hat{\mathbf{r}}_{ij}\;\hat{\mathbf{r}}_{ij}\right]\times\frac{\mathbf{q}}{m_\pi}+(i\rightleftharpoons j)\,, \tag{A.74}$$

that contains the transition form factor is defined as,

$$G_{\gamma\pi\rho}(q_\mu^2)=\frac{1}{1+q_\mu^2/m_\omega^2}\,, \tag{A.75}$$

where m_ω is the ω-meson mass, an IV term composed by two terms proportional to the d_8' and d_{21}' LECs, respectively:

$$\mathbf{j}^{\rm N3LO}_{\rm IV\;OPE}(\mathbf{q})=\mathbf{j}^{{\rm N3LO},\,d_8'}_{\rm IV\;OPE}(\mathbf{q})+\mathbf{j}^{{\rm N3LO},\,d_{21}'}_{\rm IV\;OPE}(\mathbf{q})\,. \tag{A.76}$$

These currents are defined as

$$\mathbf{j}^{(1),\,d_8'}_{\rm IV\;OPE}(\mathbf{q})\;=\;-i\,\mathrm{e}^{i\,\mathbf{q}\cdot\mathbf{r}_i}\,\frac{G_{\gamma N\Delta}(q_\mu^2)}{\mu_{\gamma N\Delta}}\,\tau_{j,z}\left[I_1^V(x_{ij})\,\boldsymbol{\sigma}_j+I_2^V(x_{ij})\;\boldsymbol{\sigma}_j\cdot\hat{\mathbf{r}}_{ij}\;\hat{\mathbf{r}}_{ij}\right]\times\frac{\mathbf{q}}{m_\pi}+(i\rightleftharpoons j), \tag{A.77}$$

$$\mathbf{j}^{(1),\,d_{21}'}_{\rm IV\;OPE}(\mathbf{q})\;=\;\frac{i}{4}\,\mathrm{e}^{i\,\mathbf{q}\cdot\mathbf{r}_i}\,\frac{G_{\gamma N\Delta}(q_\mu^2)}{\mu_{\gamma N\Delta}}\,\left(\boldsymbol{\tau}_i\times\boldsymbol{\tau}_j\right)_z\left[\widetilde{I}_1^V(x_{ij})\,\boldsymbol{\sigma}_i\times\boldsymbol{\sigma}_j+\widetilde{I}_2^V(x_{ij})\,\boldsymbol{\sigma}_j\cdot\hat{\mathbf{r}}_{ij}\;\boldsymbol{\sigma}_i\times\hat{\mathbf{r}}_{ij}\right]\times\frac{\mathbf{q}}{m_\pi}$$
$$+(i\rightleftharpoons j)\,. \tag{A.78}$$

In addition to the one-pion range contribution, there are two-pion loops,

$$\mathbf{j}^{(1)}_{\rm LOOP} = i\,\tau_{j,z}\,{\rm e}^{i\mathbf{q}\cdot\mathbf{R}_{ij}}\,G_E^V(q_\mu^2)\left\{\left(\left[F_0^{(0)}(\lambda_{ij}) + F_2^{(1)}(\lambda_{ij})\right]\boldsymbol{\sigma}_i + F_2^{(2)}(\lambda_{ij})\,\boldsymbol{\sigma}_i\cdot\hat{\mathbf{r}}_{ij}\,\hat{\mathbf{r}}_{ij}\right)\times\frac{\mathbf{q}}{2\,m_\pi}\right.$$
$$\left.-\frac{1}{2}(\boldsymbol{\tau}_i\times\boldsymbol{\tau}_j)_z\,{\rm e}^{i\mathbf{q}\cdot\mathbf{R}_{ij}}\,\hat{\mathbf{r}}_{ij}\left[\lambda_{ij}\,\frac{\tilde{v}_{2\pi}(\lambda_{ij})}{2\,m_\pi}\right]\right\} + (i\rightleftharpoons j)\,. \tag{A.79}$$

A detailed derivation of the loop functions $F^{(n)}(\lambda_{ij})$, where $\lambda_{ij} = 2m_\pi r_{ij}$, can be found in Ref. [13]. Here, I limit myself to listing their expressions:

$$\begin{aligned}
F_0^{(0)}(\lambda) &= \frac{g_A^4}{256\,\pi^3}\,\frac{(2\,m_\pi)^4}{f_\pi^4}\left\{\frac{{\rm e}^{-\lambda}}{\lambda}\left[1+\frac{1}{2}\int_0^1 dz\,\frac{1-{\rm e}^{-\lambda(\alpha_z-1)}}{\alpha_z^2-1}\right]-\left[\left(\frac{1}{g_A^2}+1\right)\frac{1}{\lambda}-\frac{1}{2}\right]E_1^{(-)}(\lambda)\right.\\
&\quad\left.-\left[\left(\frac{1}{g_A^2}+1\right)\frac{1}{\lambda}+\frac{1}{2}\right]E_1^{(+)}(\lambda)\right\},\\
F_1^{(2)}(\lambda) &= \frac{g_A^4}{256\,\pi^3}\,\frac{(2\,m_\pi)^4}{f_\pi^4}\left\{{\rm e}^{-\lambda}\left(\frac{1}{\lambda^3}+\frac{1}{\lambda^2}\right)\left[1+\frac{1}{2}\int_0^1 dz\,\frac{1-{\rm e}^{-\lambda(\alpha_z-1)}}{\alpha_z^2-1}\right]\right.\\
&\quad-\left[\left(\frac{1}{g_A^2}+2\right)\left(\frac{1}{\lambda^3}+\frac{1}{\lambda^2}\right)-\frac{1}{2\,\lambda}\right]E_1^{(-)}(\lambda)-\left[\left(\frac{1}{g_A^2}+2\right)\left(\frac{1}{\lambda^3}-\frac{1}{\lambda^2}\right)-\frac{1}{2\,\lambda}\right]E_1^{(+)}(\lambda)\\
&\quad\left.+\left(\frac{1}{g_A^2}-2\right)\frac{1}{\lambda^3}\,E_1^{(0)}(\lambda)-\int_0^1 dz\left(\frac{4}{\lambda^3}+\frac{1}{\alpha_z+1}\,\frac{1}{2\,\lambda^2}\right){\rm e}^{-\lambda\alpha_z}\right\},\\
F_2^{(2)}(\lambda) &= -\frac{g_A^4}{256\,\pi^3}\,\frac{(2\,m_\pi)^4}{f_\pi^4}\left\{{\rm e}^{-\lambda}\left(\frac{3}{\lambda^3}+\frac{3}{\lambda^2}+\frac{1}{\lambda}\right)\left[1+\frac{1}{2}\int_0^1 dz\,\frac{1-{\rm e}^{-\lambda(\alpha_z-1)}}{\alpha_z^2-1}\right]\right.\\
&\quad-\left[\left(\frac{1}{g_A^2}+2\right)\left(\frac{3}{\lambda^3}+\frac{3}{\lambda^2}+\frac{1}{\lambda}\right)-\frac{1}{2}\left(1+\frac{1}{\lambda}\right)\right]E_1^{(-)}(\lambda)\\
&\quad-\left[\left(\frac{1}{g_A^2}+2\right)\left(\frac{3}{\lambda^3}-\frac{3}{\lambda^2}+\frac{1}{\lambda}\right)+\frac{1}{2}\left(1-\frac{1}{\lambda}\right)\right]E_1^{(+)}(\lambda)+\left(\frac{1}{g_A^2}-2\right)\frac{3}{\lambda^3}\,E_1^{(0)}(\lambda)\\
&\quad\left.-\int_0^1 dz\left[\frac{16}{\lambda^3}+\left(4\,\alpha_z+\frac{3}{2}\,\frac{1}{\alpha_z+1}\right)\frac{1}{\lambda^2}\right]{\rm e}^{-\lambda\alpha_z}\right\},
\end{aligned}$$

where the following definitions have been employed,

$$\alpha_z = (1-z^2)^{-1/2}\,, \tag{A.80}$$

$$E_1(x) = \int_x^\infty dt\frac{e^{-t}}{t}\,, \tag{A.81}$$

$$E_1^{(\pm)}(\lambda) = \frac{e^{\pm\lambda}}{2}\int_0^1 dz E_1(\lambda\alpha_z+\lambda) \tag{A.82}$$

$$E_1^{(0)}(\lambda) = \int_0^1 dz E_1(\lambda\alpha_z) \tag{A.83}$$

The N3LO LOOP two-body currents satisfy the continuity equation with the NV2 potential at N2LO in the limit of momentum transfer going to zero. This is reflected by the presence of the $\tilde{v}_{2\pi}(\lambda_{ij}) = v_{2\pi}(\lambda_{ij})\boldsymbol{\tau}_i\cdot\boldsymbol{\tau}_j$ term in Eq. (A.79). The $v_{2\pi}(\lambda_{ij})$ term indicates contributions to the NV2 potential coming from TPE diagrams. In the present study, it accounts for TPE terms at both NLO and N2LO, therefore it includes two-pion loops with and without Δ-isobar intermediate states (see Ref. [13] for more details).

The correlation functions in the IS term may be obtained from the following rescaling of the N2LO-Δ correlation functions,

$$\left(\frac{\tilde{g}_A h_A}{18\pi}\frac{\mu_{\Delta N}}{2m}\frac{m_\pi^2}{m_{\Delta N}}\frac{m_\pi^2}{f_\pi^2}\right) \longrightarrow \left(\frac{\tilde{g}_A}{16\pi}\frac{m_\pi^2}{f_\pi^2} m_\pi^2\, d_9'\right) , \tag{A.84}$$

where $\mu_{\Delta N} = 3$ n. m. is the N-to-Δ magnetic transition magnetic moment [15]. Similarly, the IV correlation function $I_1^V(\mu_{ij})$ is obtained by substituting,

$$\left(\frac{\tilde{g}_A h_A}{18\pi}\frac{\mu_{\Delta N}}{2m}\frac{m_\pi^2}{m_{\Delta N}}\frac{m_\pi^2}{f_\pi^2}\right) \longrightarrow \left(\frac{\tilde{g}_A}{16\pi}\frac{m_\pi^2}{f_\pi^2} m_\pi^2\, d_8'\right) . \tag{A.85}$$

while the function $\widetilde{I}_1^V(\mu_{ij})$ can be obtained by making the substitution,

$$\left(\frac{\tilde{g}_A h_A}{18\pi}\frac{\mu_{\Delta N}}{2m}\frac{m_\pi^2}{m_{\Delta N}}\frac{m_\pi^2}{f_\pi^2}\right) \longrightarrow \left(\frac{\tilde{g}_A}{16\pi}\frac{m_\pi^2}{f_\pi^2} m_\pi^2\, d_{21}'\right) . \tag{A.86}$$

The LECs d_8', d_9', and d_{21}' are unknown and must be fit to data. Additionally, there are short-range terms which can be divided into an IS and IV term obtained via minimal substitution,

$$\mathbf{j}^{(1)}_{\rm MIN\ IS}(\mathbf{q}) = -\frac{1}{8} m_\pi^4\, C_5\, {\rm e}^{i\mathbf{q}\cdot\mathbf{R}_{ij}} G_E^S(q_\mu^2)\Big[i\, C^{(0)}_{R_S}(z_{ij})\, (\boldsymbol{\sigma}_i + \boldsymbol{\sigma}_j) \times \frac{\mathbf{q}}{m_\pi}\Big] , \tag{A.87}$$

$$\begin{aligned}\mathbf{j}^{(1)}_{\rm MIN\ IV}(\mathbf{q}) = \frac{1}{8}\ &(\boldsymbol{\tau}_i \times \boldsymbol{\tau}_j)_z\ {\rm e}^{i\mathbf{q}\cdot\mathbf{R}_{ij}} G_E^V(q_\mu^2)\, C^{(1)}_{R_S}(z_{ij})\Big[m_\pi^4\,(C_2 + 3\,C_4 + C_7)\,\hat{\mathbf{r}}_{ij}\\ &+ m_\pi^4\,(C_2 - C_4 - C_7)\ \hat{\mathbf{r}}_{ij}\ \boldsymbol{\sigma}_i\cdot\boldsymbol{\sigma}_j + m_\pi^4\, C_7\,(\boldsymbol{\sigma}_i\cdot\hat{\mathbf{r}}_{ij}\ \boldsymbol{\sigma}_j + \boldsymbol{\sigma}_j\cdot\hat{\mathbf{r}}_{ij}\ \boldsymbol{\sigma}_i)\Big]\\ &-\frac{1}{8} m_\pi^4\, C_5\, {\rm e}^{i\mathbf{q}\cdot\mathbf{R}_{ij}}\Big[\,(\tau_{i,z} - \tau_{j,z})\ C^{(1)}_{R_S}(z_{ij})\,(\boldsymbol{\sigma}_i + \boldsymbol{\sigma}_j)\times\hat{\mathbf{r}}_{ij}\Big] .\end{aligned} \tag{A.88}$$

a two non-minimal terms,

$$\mathbf{j}^{(1)}_{\rm NM\,IS}(\mathbf{q}) = -i\, {\rm e}^{i\mathbf{q}\cdot\mathbf{R}_{ij}}\, C^{(0)}_{R_S}(z_{ij})\, m_\pi^4\, G_E^S(q_\mu^2)\, C_{15}'\,(\boldsymbol{\sigma}_i + \boldsymbol{\sigma}_j)\times\frac{\mathbf{q}}{m_\pi} , \tag{A.89}$$

$$\mathbf{j}^{(1)}_{\rm NM\,IV}(\mathbf{q}) = -i\, {\rm e}^{i\mathbf{q}\cdot\mathbf{R}_{ij}}\, C^{(0)}_{R_S}(z_{ij})\, m_\pi^4\, G_E^V(q_\mu^2)\, C_{16}'\,(\tau_{i,z} - \tau_{j,z})\,(\boldsymbol{\sigma}_i - \boldsymbol{\sigma}_j)\times\frac{\mathbf{q}}{m_\pi} . \tag{A.90}$$

and a term needed to regularize the N3LO loop contribution,

$$\mathbf{j}^{\rm N3LO}_{\rm CT}(\mathbf{q}) = i\,\tau_{j,z}\ {\rm e}^{i\mathbf{q}\cdot\mathbf{R}_{ij}}\ F_0^{(0)}(z_{ij};\infty)\ \boldsymbol{\sigma}_i\times\frac{\mathbf{q}}{2\,m_\pi} + (i \rightleftharpoons j) . \tag{A.91}$$

Table A.6 LECs entering the vector current at N3LO fit in Ref. [14]. Note that these LECs have been fit only for a subset of NV2+3 models; namely, to the models with the NV3 constrained by weak interaction data only

Model	d_1^V	d_2^V	d_3^V	d_1^S	d_2^S
Ia*	−0.050(2)	0.49(7)	0.094(4)	0.012(1)	0.023(7)
IIa*	−0.066(2)	0.01(7)	0.069(4)	0.011(1)	0.019(7)
Ib*	−0.055(3)	0.09(5)	0.073(3)	0.025(2)	0.030(6)
IIb*	−0.049(2)	0.09(4)	0.048(3)	0.017(1)	0.018(3)

In the equations above

$$C_{R_S}^{(0)} = \frac{e^{-z^2}}{\pi^{3/2}(m_\pi R_S)^3}, \tag{A.92}$$

$$C_{R_S}^{(1)} = \frac{1}{m_\pi R_S}\frac{dC_{R_S}^{(0)}}{dz}, \tag{A.93}$$

where $z_{ij} = r_{ij}/R_S$.

The LECs C_i in Eqs. A.87 and A.88 are the same that appear in the short-range part of the NV2+3, while the LECs C'_{15} and C'_{16} are unknown and must be fit to data. When fitting in Ref. [14], the following dimensionless quantities were used,

$$d_1^S = m_\pi^4 C'_{15} \tag{A.94}$$

$$d_2^S = m_\pi^2 d'_9 \tag{A.95}$$

$$d_1^V = m_\pi^4 C'_{16} \tag{A.96}$$

$$d_2^V = m_\pi^2 d'_8 \tag{A.97}$$

$$d_3^V = m_\pi^2 d'_{21}, \tag{A.98}$$

and the values of these constants are reported in Table A.6.

A.3 Tree-Level Axial Charge Operator

The axial currents consistent with the NV2+3 were first reported in Ref. [16] and I summarize the contributions in this chapter. At LO (Q^{-2}), the charge reads,

$$\rho_{5,a}^{(-1)} = -\frac{g_A(q_\mu^2)}{2}\tau_{i,a}\left(\frac{1}{m_N}\boldsymbol{\sigma}_i\cdot\mathbf{K}_i - \frac{\omega}{\mathbf{q}^2+m_\pi^2}\boldsymbol{\sigma}_i\cdot\mathbf{q}\right)(2\pi)^3\delta(\mathbf{k_i}-\mathbf{q}), \tag{A.99}$$

which includes the previously missing pseudoscalar term proportional to the energy transfer ω that was first reported in Ref. [17]. The axial form factor adopted in this work has the form,

$$g_A(q_\mu^2) = (1 + q_\mu^2/\Lambda_A^2)^{-2}\,, \tag{A.100}$$

with $q_\mu^2 = \mathbf{q}^2 - \omega^2$, as for the vector current, and $\Lambda_A = 1.05$ GeV adopted in this thesis. In what remains of this section, I will suppress the q_μ^2 dependence on the form factor, but it is implied. Note that the value of $g_A(0) = 1.2723$ in the axial current is taken as the value in the 2018 PDG [18] (which was adopted for the calculations in this thesis) differs from that used in the nuclear potentials and the vector current operators, which account for the Goldberger-Treiman discrepancy [3, 4]. In the computation of the β-decay spectrum of ^{6}He in Sect. 4.3, the more recent $g_A = 1.2754$ [19] was used. The subscript a denotes the isospin component of the current while i labels the nucleon. At tree-level, to the order considered in the work of this thesis, the only subleading contribution is from an OPE diagram at Q^{-1},

$$\rho_{5,a}^{(-1)} = i\,\frac{g_A}{4f_\pi^2}\,(\boldsymbol{\tau}_i \times \boldsymbol{\tau}_j)_a\,\boldsymbol{\sigma}_j\cdot\mathbf{k}_j\,\frac{1}{\omega_j^2}. \tag{A.101}$$

A.4 Tree-Level Axial Current Operator

The axial current operator was first derived in Ref. [16] without Δ-isobars and the Δ-full expressions were presented first in Ref. [11]. At LO (Q^{-3}), the current is given by,

$$\mathbf{j}_{5,a}^{(-3)}(\mathbf{q}) = -\frac{g_A(q_\mu^2)}{2}\,\tau_{i,a}\,\boldsymbol{\sigma}_i\,\mathrm{e}^{i\mathbf{q}\cdot\mathbf{r}_i}\,, \tag{A.102}$$

where a is again denoting the isospin index. As for the charge, I will suppress the q_μ^2 dependence in the remainder of this section but it is implied. The current has no NLO contribution. At N2LO (Q^{-1}), there is a one-body relativististic term,

$$\begin{aligned}\mathbf{j}_{5,a}^{(-1)}(\mathbf{q};\mathrm{RC}) &= \frac{g_A}{8\,m^2}\,\tau_{i,a}\left[\mathbf{p}_i\times(\boldsymbol{\sigma}_i\times\mathbf{p}_i)\;,\;\mathrm{e}^{i\mathbf{q}\cdot\mathbf{r}_i}\right]_+\\ &+\frac{g_A}{8\,m^2}\,\tau_{i,a}\,\mathrm{e}^{i\mathbf{q}\cdot\mathbf{r}_i}\,(i\,\mathbf{q}\times\mathbf{p}_i+\mathbf{q}\,\boldsymbol{\sigma}_i\cdot\mathbf{q}/2)\,,\end{aligned} \tag{A.103}$$

plus a two-body one-pion range operator involving the exchange of an intermediate Δ,

$$\begin{aligned}\mathbf{j}_{5,a}^{\mathrm{N2LO}}(\mathbf{q};\Delta) &= -\mathrm{e}^{i\,\mathbf{q}\cdot\mathbf{r}_i}\,(\boldsymbol{\tau}_i\times\boldsymbol{\tau}_j)_a\left[I^{(1)}(\mu_{ij};\alpha_1^\Delta)\,\boldsymbol{\sigma}_i\times\boldsymbol{\sigma}_j+I^{(2)}(\mu_{ij};\alpha_1^\Delta)\,\boldsymbol{\sigma}_i\times\hat{\mathbf{r}}_{ij}\,\boldsymbol{\sigma}_j\cdot\hat{\mathbf{r}}_{ij}\right]\\ &-\mathrm{e}^{i\mathbf{q}\cdot\mathbf{r}_i}\,\tau_{j,a}\left[I^{(1)}(\mu_{ij};\alpha_2^\Delta)\,\boldsymbol{\sigma}_j+I^{(2)}(\mu_{ij};\alpha_2^\Delta)\,\hat{\mathbf{r}}_{ij}\,\boldsymbol{\sigma}_j\cdot\hat{\mathbf{r}}_{ij}\right]+(i\rightleftharpoons j)\,,\end{aligned} \tag{A.104}$$

At N3LO (Q^0), there is an OPE contribution,

$$\begin{aligned}\mathbf{j}_{5,a}^{\rm N3LO}(\mathbf{q};\,{\rm OPE}) &= -{\rm e}^{i\,\mathbf{q}\cdot\mathbf{r}_i}\,\left(\boldsymbol{\tau}_i\times\boldsymbol{\tau}_j\right)_a\left[I^{(1)}(\mu_{ij};\alpha_1)\,\boldsymbol{\sigma}_i\times\boldsymbol{\sigma}_j+I^{(2)}(\mu_{ij};\alpha_1)\,\boldsymbol{\sigma}_i\times\hat{\mathbf{r}}_{ij}\,\boldsymbol{\sigma}_j\cdot\hat{\mathbf{r}}_{ij}\right]\\
&\quad-{\rm e}^{i\mathbf{q}\cdot\mathbf{r}_i}\,\tau_{j,a}\left[I^{(1)}(\mu_{ij};\alpha_2)\,\boldsymbol{\sigma}_j+I^{(2)}(\mu_{ij};\alpha_2)\,\hat{\mathbf{r}}_{ij}\,\boldsymbol{\sigma}_j\cdot\hat{\mathbf{r}}_{ij}\right]\\
&\quad-\left(\boldsymbol{\tau}_i\times\boldsymbol{\tau}_j\right)_a\frac{1}{2\,m_\pi}\left\{\mathbf{p}_i\;,\;{\rm e}^{i\mathbf{q}\cdot\mathbf{r}_i}\,\widetilde{I}^{(1)}(\mu_{ij};\widetilde{\alpha}_1)\,\boldsymbol{\sigma}_j\cdot\hat{\mathbf{r}}_{ij}\right\}\\
&\quad-i\,\left(\boldsymbol{\tau}_i\times\boldsymbol{\tau}_j\right)_a{\rm e}^{i\mathbf{q}\cdot\mathbf{r}_i}\,\widetilde{I}^{(1)}(\mu_{ij};\widetilde{\alpha}_2)\,\boldsymbol{\sigma}_i\times\frac{\mathbf{q}}{m_\pi}\,\boldsymbol{\sigma}_j\cdot\hat{\mathbf{r}}_{ij}+(i\rightleftharpoons j)\,,\qquad\text{(A.105)}\end{aligned}$$

plus a CT term,

$$\mathbf{j}_{5,a}^{\rm N3LO}(\mathbf{q};\,{\rm CT})=z_0\,{\rm e}^{i\,\mathbf{q}\cdot\mathbf{R}_{ij}}\,\frac{{\rm e}^{-z_{ij}^2}}{\pi^{3/2}}\,\left(\boldsymbol{\tau}_i\times\boldsymbol{\tau}_j\right)_a\,\left(\boldsymbol{\sigma}_i\times\boldsymbol{\sigma}_j\right)\,,\qquad\text{(A.106)}$$

with the parameter z_0 governing the strength of this term is given by the following combination of LECs and model parameters,

$$z_0=\frac{g_A}{2}\,\frac{m_\pi^2}{f_\pi^2}\,\frac{1}{(m_\pi\,R_{\rm S})^3}\left[-\frac{m_\pi}{4\,g_A\,\Lambda_\chi}\,c_D+\frac{m_\pi}{3}\,(c_3+2\,c_4)+\frac{m_\pi}{6\,m}\right],\qquad\text{(A.107)}$$

where c_3 and c_4 are LECs from the πN Lagrangian and c_D is the same LEC that appears in the three-body force contact contribution. Finally, the correlation functions are given by,

$$I^{(1)}(\mu;\alpha)=-\alpha\,(1+\mu)\,\frac{e^{-\mu}}{\mu^3}\,,\qquad\text{(A.108)}$$

$$I^{(2)}(\mu;\alpha)=\alpha\,(3+3\,\mu+\mu^2)\,\frac{e^{-\mu}}{\mu^3}\,,\qquad\text{(A.109)}$$

$$\widetilde{I}^{(1)}(\mu;\widetilde{\alpha})=-\widetilde{\alpha}\,(1+\mu)\,\frac{e^{-\mu}}{\mu^2}\,,\qquad\text{(A.110)}$$

where

$$\alpha_1^\Delta=\frac{g_A}{8\,\pi}\,\frac{m_\pi^3}{f_\pi^2}\,c_4^\Delta\qquad\text{(A.111)}$$

$$\alpha_2^\Delta=\frac{g_A}{4\,\pi}\,\frac{m_\pi^3}{f_\pi^2}\,c_3^\Delta\qquad\text{(A.112)}$$

$$\alpha_1=\frac{g_A}{8\,\pi}\,\frac{m_\pi^3}{f_\pi^2}\left(c_4+\frac{1}{4\,m}\right)\qquad\text{(A.113)}$$

$$\alpha_2=\frac{g_A}{4\,\pi}\,\frac{m_\pi^3}{f_\pi^2}\,c_3\qquad\text{(A.114)}$$

$$\widetilde{\alpha}_1=\frac{g_A}{16\,\pi}\,\frac{m_\pi^3}{m\,f_\pi^2}\qquad\text{(A.115)}$$

$$\widetilde{\alpha}_2 = \frac{g_A}{32\,\pi}\,\frac{m_\pi^3}{m\,f_\pi^2}\,(c_6+1)\;, \tag{A.116}$$

with the LECs c_3^Δ and c_4^Δ defined as [11],

$$c_3^\Delta = -\frac{h_A^2}{9\,m_{\Delta N}} \tag{A.117}$$

$$c_4^\Delta = \frac{h_A^2}{18\,m_{\Delta N}}\;. \tag{A.118}$$

References

1. S. Pastore, L. Girlanda, R. Schiavilla, and M. Viviani. The two-nucleon electromagnetic charge operator in chiral effective field theory (χEFT) up to one loop. *Phys. Rev.*, C84:024001, 2011. doi: 10.1103/PhysRevC.84.024001.
2. M. Piarulli, L. Girlanda, R. Schiavilla, R. Navarro Pérez, J. E. Amaro, and E. Ruiz Arriola. Minimally nonlocal nucleon-nucleon potentials with chiral two-pion exchange including Δ resonances. *Phys. Rev.*, C91(2):024003, 2015. doi: 10.1103/PhysRevC.91.024003.
3. R. A. Arndt, R. L. Workman, and M. M. Pavan. Pion nucleon partial wave analysis with fixed t dispersion relation constraints. *Phys. Rev.*, C49:2729–2734, 1994. doi: 10.1103/PhysRevC.49.2729.
4. Vincent G. J. Stoks, Rob Timmermans, and J. J. de Swart. On the pion - nucleon coupling constant. *Phys. Rev.*, C47:512–520, 1993. doi: 10.1103/PhysRevC.47.512.
5. Norbert Kaiser, R. Brockmann, and W. Weise. Peripheral nucleon-nucleon phase shifts and chiral symmetry. *Nucl. Phys.*, A625:758–788, 1997. doi: 10.1016/S0375-9474(97)00586-1.
6. Maria Piarulli, Luca Girlanda, Rocco Schiavilla, Alejandro Kievsky, Alessandro Lovato, Laura E. Marcucci, Steven C. Pieper, Michele Viviani, and Robert B. Wiringa. Local chiral potentials with Δ-intermediate states and the structure of light nuclei. *Phys. Rev.*, C94(5):054007, 2016. doi: 10.1103/PhysRevC.94.054007.
7. Robert B. Wiringa, V. G. J. Stoks, and R. Schiavilla. An Accurate nucleon-nucleon potential with charge independence breaking. *Phys. Rev.*, C51:38–51, 1995. doi: 10.1103/PhysRevC.51.38.
8. V. E. Krohn and G. R. Ringo. Reconsiderations of the electron - neutron scattering length as measured by the scattering of thermal neutrons by noble gases. *Phys. Rev. D*, 8:1305–1307, 1973. doi: 10.1103/PhysRevD.8.1305.
9. M. Piarulli et al. Light-nuclei spectra from chiral dynamics. *Phys. Rev. Lett.*, 120(5):052503, 2018. doi: 10.1103/PhysRevLett.120.052503.
10. Nadia Fettes, Ulf-G. Meissner, Martin Mojzis, and Sven Steininger. The Chiral effective pion nucleon Lagrangian of order p**4. *Annals Phys.*, 283:273–302, 2000. doi: 10.1006/aphy.2000.6059. [Erratum: Annals Phys.288,249(2001)].
11. A. Baroni et al. Local chiral interactions, the tritium Gamow-Teller matrix element, and the three-nucleon contact term. *Phys. Rev.*, C98(4):044003, 2018. doi: 10.1103/PhysRevC.98.044003.
12. S. Pastore, L. Girlanda, R. Schiavilla, M. Viviani, and R. B. Wiringa. Electromagnetic Currents and Magnetic Moments in (chi)EFT. *Phys. Rev.*, C80:034004, 2009. doi: 10.1103/PhysRevC.80.034004.
13. R. Schiavilla, A. Baroni, S. Pastore, M. Piarulli, L. Girlanda, A. Kievsky, A. Lovato, L. E. Marcucci, Steven C. Pieper, M. Viviani, and R. B. Wiringa. Local chiral interactions and magnetic

structure of few-nucleon systems. *Phys. Rev. C*, 99:034005, Mar 2019. doi: 10.1103/PhysRevC.99.034005. URL https://link.aps.org/doi/10.1103/PhysRevC.99.034005.
14. Alex Gnech and Rocco Schiavilla. Magnetic structure of few-nucleon systems at high momentum transfers in a chiral effective field theory approach. *Phys. Rev. C*, 106(4):044001, 2022. doi: 10.1103/PhysRevC.106.044001.
15. Carl E. Carlson. ELECTROMAGNETIC N - DELTA TRANSITION AT HIGH Q**2. *Phys. Rev.*, D34:2704, 1986. doi: 10.1103/PhysRevD.34.2704.
16. A. Baroni, L. Girlanda, S. Pastore, R. Schiavilla, and M. Viviani. Nuclear Axial Currents in Chiral Effective Field Theory. *Phys. Rev.*, C93(1):015501, 2016. doi: 10.1103/PhysRevC.93.049902, 10.1103/PhysRevC.93.015501, 10.1103/PhysRevC.95.059901. [Erratum: Phys. Rev.C95,no.5,059901(2017)].
17. Garrett B. King, Alessandro Baroni, Vincenzo Cirigliano, Stefano Gandolfi, Leendert Hayen, Emanuele Mereghetti, Saori Pastore, and Maria Piarulli. Ab initio calculation of the β-decay spectrum of He6. *Phys. Rev. C*, 107(1):015503, 2023. doi: 10.1103/PhysRevC.107.015503.
18. M. Tanabashi, K. Hagiwara, K. Hikasa, K. Nakamura, Y. Sumino, F. Takahashi, J. Tanaka, K. Agashe, G. Aielli, C. Amsler, M. Antonelli, D. M. Asner, H. Baer, Sw. Banerjee, R. M. Barnett, T. Basaglia, C. W. Bauer, J. J. Beatty, V. I. Belousov, J. Beringer, S. Bethke, A. Bettini, H. Bichsel, O. Biebel, K. M. Black, E. Blucher, O. Buchmuller, V. Burkert, M. A. Bychkov, R. N. Cahn, M. Carena, A. Ceccucci, A. Cerri, D. Chakraborty, M.-C. Chen, R. S. Chivukula, G. Cowan, O. Dahl, G. D'Ambrosio, T. Damour, D. de Florian, A. de Gouvêa, T. DeGrand, P. de Jong, G. Dissertori, B. A. Dobrescu, M. D'Onofrio, M. Doser, M. Drees, H. K. Dreiner, D. A. Dwyer, P. Eerola, S. Eidelman, J. Ellis, J. Erler, V. V. Ezhela, W. Fetscher, B. D. Fields, R. Firestone, B. Foster, A. Freitas, H. Gallagher, L. Garren, H.-J. Gerber, G. Gerbier, T. Gershon, Y. Gershtein, T. Gherghetta, A. A. Godizov, M. Goodman, C. Grab, A. V. Gritsan, C. Grojean, D. E. Groom, M. Grünewald, A. Gurtu, T. Gutsche, H. E. Haber, C. Hanhart, S. Hashimoto, Y. Hayato, K. G. Hayes, A. Hebecker, S. Heinemeyer, B. Heltsley, J. J. Hernández-Rey, J. Hisano, A. Höcker, J. Holder, A. Holtkamp, T. Hyodo, K. D. Irwin, K. F. Johnson, M. Kado, M. Karliner, U. F. Katz, S. R. Klein, E. Klempt, R. V. Kowalewski, F. Krauss, M. Kreps, B. Krusche, Yu. V. Kuyanov, Y. Kwon, O. Lahav, J. Laiho, J. Lesgourgues, A. Liddle, Z. Ligeti, C.-J. Lin, C. Lippmann, T. M. Liss, L. Littenberg, K. S. Lugovsky, S. B. Lugovsky, A. Lusiani, Y. Makida, F. Maltoni, T. Mannel, A. V. Manohar, W. J. Marciano, A. D. Martin, A. Masoni, J. Matthews, U.-G. Meißner, D. Milstead, R. E. Mitchell, K. Mönig, P. Molaro, F. Moortgat, M. Moskovic, H. Murayama, M. Narain, P. Nason, S. Navas, M. Neubert, P. Nevski, Y. Nir, K. A. Olive, S. Pagan Griso, J. Parsons, C. Patrignani, J. A. Peacock, M. Pennington, S. T. Petcov, V. A. Petrov, E. Pianori, A. Piepke, A. Pomarol, A. Quadt, J. Rademacker, G. Raffelt, B. N. Ratcliff, P. Richardson, A. Ringwald, S. Roesler, S. Rolli, A. Romaniouk, L. J. Rosenberg, J. L. Rosner, G. Rybka, R. A. Ryutin, C. T. Sachrajda, Y. Sakai, G. P. Salam, S. Sarkar, F. Sauli, O. Schneider, K. Scholberg, A. J. Schwartz, D. Scott, V. Sharma, S. R. Sharpe, T. Shutt, M. Silari, T. Sjöstrand, P. Skands, T. Skwarnicki, J. G. Smith, G. F. Smoot, S. Spanier, H. Spieler, C. Spiering, A. Stahl, S. L. Stone, T. Sumiyoshi, M. J. Syphers, K. Terashi, J. Terning, U. Thoma, R. S. Thorne, L. Tiator, M. Titov, N. P. Tkachenko, N. A. Törnqvist, D. R. Tovey, G. Valencia, R. Van de Water, N. Varelas, G. Venanzoni, L. Verde, M. G. Vincter, P. Vogel, A. Vogt, S. P. Wakely, W. Walkowiak, C. W. Walter, D. Wands, D. R. Ward, M. O. Wascko, G. Weiglein, D. H. Weinberg, E. J. Weinberg, M. White, L. R. Wiencke, S. Willocq, C. G. Wohl, J. Womersley, C. L. Woody, R. L. Workman, W.-M. Yao, G. P. Zeller, O. V. Zenin, R.-Y. Zhu, S.-L. Zhu, F. Zimmermann, P. A. Zyla, J. Anderson, L. Fuller, V. S. Lugovsky, and P. Schaffner. Review of particle physics. *Phys. Rev. D*, 98:030001, Aug 2018. doi: 10.1103/PhysRevD.98.030001. URL https://link.aps.org/doi/10.1103/PhysRevD.98.030001.
19. P.A. Zyla et al. Review of Particle Physics. *PTEP*, 2020(8):083C01, 2020. doi: 10.1093/ptep/ptaa104.

Appendix B
The Multipole Decomposition and Nuclear Matrix Elements

The typical approach to the study of nuclear electroweak processes is the standard multipole analysis where the perturbing Hamiltonian may be broken down into operators of different multipolarities whose summed contribution gives the matrix element of interest. In many-body approaches, it is easier to compute matrix elements of charge and current operators directly. In this chapter, I discuss the relationship between the Fourier transform of the coordinate space currents and the standard multipole moments.

First, I note that there is a detailed derivation of the standard multipole operators in Ref. [1]. For the purposes of this thesis, I will not rework that derivation and direct interested readers to the aforementioned reference. Instead, I will take the result as defining of the multipoles for an external field transferring momentum $\mathbf{q}$ to the nucleus,

$$C_{JM}(q) = \int d^3x [j_J(qx) Y_{JM}(\hat{\mathbf{x}})] \rho(\mathbf{x}) \tag{B.1}$$

$$L_{JM}(q) = \frac{i}{q} \int d^3x \{\nabla [j_J(qx) Y_{JM}(\hat{\mathbf{x}})]\} \cdot \mathbf{j}(\mathbf{x}) \tag{B.2}$$

$$E_{JM}(q) = \frac{1}{q} \int d^3x [\nabla \times j_J(qx) \mathbf{Y}^M_{JJ1}(\hat{\mathbf{x}})] \cdot \mathbf{j}(\mathbf{x}) \tag{B.3}$$

$$M_{JM}(q) = \int d^3x [j_j(qx) \mathbf{Y}^M_{JJ1}(\hat{\mathbf{x}}) \cdot \mathbf{j}(\mathbf{x}) \,, \tag{B.4}$$

where $j_J(qx)$ are spherical Bessel functions, $Y_{JM}(\hat{\mathbf{x}})$ are the spherical harmonics, and $\mathbf{Y}^M_{JJ1}(\hat{\mathbf{x}})$ are vector spherical harmonics [2]. At this point, I have made no assumptions about the specific transformation properties of the charge and current operators. I have only assumed that some local charge and current density coupling the nucleus to the external field exist.

G. B. King, *Electroweak Structure and Reactions in Light Nuclei with Quantum Monte Carlo Methods*, Springer Theses, https://doi.org/10.1007/978-3-032-03191-4

I define the spherical polarizations of the external (axial-)vector field as,

$$\hat{\mathbf{e}}_0 = \frac{\mathbf{q}}{q} \tag{B.5}$$

$$\hat{\mathbf{e}}_{\pm 1} = \mp\frac{1}{\sqrt{2}}\left(\hat{\mathbf{e}}_1 + \hat{\mathbf{e}}_2\right) , \tag{B.6}$$

where the unit vectors $\hat{\mathbf{e}}_1$ and $\hat{\mathbf{e}}_2$ define the directions transverse to $\hat{\mathbf{e}}_0$ to form an orthonormal basis. The spherical polarizations also form an orthonormal basis and may be used to represent the current operator in terms of components that are longitudinal and transverse to $\mathbf{q}$. The definitions in Eqs. B.1 through B.4 may be used to express the charge operator, as well as the spherical components of the current operators in momentum space [3, 4],

$$\rho(\mathbf{q}) = \sum_{LM_L} 4\pi i^L Y^*_{LM_L}(\hat{\mathbf{q}}) C_{LM_L}(q) \tag{B.7}$$

$$\mathbf{j}_0(\mathbf{q}) = -\sum_{LM_L} 4\pi i^L Y^*_{LM_L}(\hat{\mathbf{q}}) L_{LM_L}(q) \tag{B.8}$$

$$\begin{aligned} \mathbf{j}_{\lambda=\pm 1}(\mathbf{q}) = -\sum_{LM_L, L\geq 1} & \sqrt{2\pi(2L+1)} i^L D^L_{M_L\lambda}(-\phi, -\theta, \phi) \\ & \times [E_{LM_L}(q) + \lambda M_{LM_L}(q)] , \end{aligned} \tag{B.9}$$

where $D^L_{M_L,\lambda}(-\phi, -\theta, \phi)$ is a Wigner D-matrix [2] and the quantities ϕ and θ define, respectively, the azimuthal and polar angles that $\mathbf{q}$ forms with respect to the spin quantization axis $\hat{\mathbf{z}}$. The rotation operator that connects states quantized along $\hat{\mathbf{e}}_0$ to states quantized along $\hat{\mathbf{z}}$ is,

$$R(-\phi, \theta, \phi) = e^{-i\phi J_z} e^{-i\theta J_y} e^{i\phi J_Z} . \tag{B.10}$$

Conventionally, one defines, [2]

$$D^L_{M',M}(-\phi, \theta, \phi) = {}_{\hat{\mathbf{e}}_0}\langle LM'| R(-\phi, \theta, \phi) |LM\rangle_{\hat{\mathbf{e}}_0} = e^{-iM'\phi} d^L_{M',M}(\theta) e^{iM\phi} , \tag{B.11}$$

such that,

$$|LM\rangle_{\hat{\mathbf{z}}} = R(-\phi, \theta, \phi) |LM\rangle_{\hat{\mathbf{e}}_0} = \sum_{M'} D^L_{M',M}(-\phi, \theta, \phi) |LM'\rangle_{\hat{\mathbf{e}}_0} . \tag{B.12}$$

Noth that the subscript on the bras and kets indicates the direction along which the state is d. Recall that $D^L_{M',M}(-\phi, \theta, \phi)$ allows one to define the behavior of an irreducible tensor operator (ITO) of rank L under rotations. Namely, such an operator transforms as [2]

$$[R(-\phi,\theta,\phi)]^\dagger T_{LM}[R(-\phi,\theta,\phi)] = [R(-\phi,-\theta,\phi)]T_{LM}[R(-\phi,-\theta,\phi)]^\dagger$$
$$= \sum_{M'} D^L_{M',M}(-\phi,-\theta,\phi)T_{LM'} \,. \tag{B.13}$$

It is easily verifiable from the transformation properties of $Y_{JM}(\hat{\mathbf{x}})$ and $\mathbf{Y}^M_{JJ1}(\hat{\mathbf{x}})$ that the definitions of the multipoles in Eqs. B.1 through B.4 are ITOs of rank J. Thus, when computing matrix elements of the current, we can leverage the Wigner-Eckart theorem, which states that matrix elements of a tensor operator of rank L with respect to angular momentum eigenstates is given by,

$$\langle J_f M_f|T_{LM}|J_f M_f\rangle = \frac{(-1)^{(J_i-M_i)}}{\sqrt{2L+1}}\langle J_f, M_f; J_i, -M_i|L, M\rangle T_L \,, \tag{B.14}$$

where the reduced multipole is $T_L = \langle J_f||T_L||J_i\rangle$ and $\langle J_f, M_f; J_i, -M_i|L, M\rangle$ is a Clebsch-Gordan coefficient.

The expressions of the charge and current operators in Eqs. B.7 through B.9 in terms of the multipole operators in Eqs. B.1 through B.4 are obtained by considering the Fourier transforms of the coordinate space operators. First, let us consider the specific case where the momentum transferred by the external field is defined along the spin-quantization axis; *i.e.*, consider the case where $\hat{\mathbf{q}} = \hat{\mathbf{e}}_0 = \hat{\mathbf{z}}$. The Fourier transform of the charge operator,

$$\rho(\mathbf{q}) = \int d\mathbf{x} e^{i\mathbf{q}\cdot\mathbf{x}}\rho(\mathbf{x}) \,, \tag{B.15}$$

may be obtained by decomposing the exponential in partial waves [2],

$$e^{i\mathbf{q}\cdot\mathbf{x}} = \sum_L i^L\sqrt{4\pi(2L+1)}\, j_L(qx)Y_{L0}\rho(\mathbf{x}) \,. \tag{B.16}$$

Then, we can rewrite Eq. B.15 as,

$$\rho(\mathbf{q}) = \int d\mathbf{x} \sum_L i^L\sqrt{4\pi(2L+1)}\, j_L(qx)Y_{L0}(\hat{\mathbf{x}})\rho(\mathbf{x}) \,. \tag{B.17}$$

Using the definition in Eq. B.1, it is clear that,

$$\rho(\mathbf{q}) = \sum_L i^L\sqrt{4\pi(2L+1)}\, C_{L0}(q) \,. \tag{B.18}$$

We may now go to case where $\hat{\mathbf{e}}_0$ defines an arbitrary direction by using the fact that $C_{LM}(q)$ is an irreducible tensor operator of rank L. Thus, using Eq. B.13, we can rewrite this expression as

$$\rho(\mathbf{q}) = \sum_{LM_L} i^L \sqrt{4\pi(2L+1)} D^L_{M_L,0}(-\phi, -\theta, \phi) C_{LM_L}(q) . \tag{B.19}$$

Now, using the property that $d^L_{M',M}(\beta) = (-1)^{M'-M} d^L_{-M,-M'}(-\beta)$ [2], we notice that

$$D^L_{M_L,0}(-\phi, -\theta, \phi) = d^L_{M_L,0}(-\theta) e^{-iM_L\phi} \tag{B.20}$$

$$= (-1)^{M_L} d_{0,-M_L}(\theta) e^{-iM_L\phi} \tag{B.21}$$

$$= (-1)^{M_L} D^L_{0,-M_L}(\phi, \theta, \phi) . \tag{B.22}$$

Finally, we can use the properties [2]

$$D^L_{0,M}(\alpha, \beta, \gamma) = \sqrt{\frac{4\pi}{2L+1}} Y_{LM}(\beta, \gamma) , \tag{B.23}$$

and

$$Y_{L,-M}(\beta, \gamma) = (-1)^M Y^*_{LM}(\beta, \gamma) , \tag{B.24}$$

to show

$$D^L_{M_L,0}(-\phi, -\theta, \phi) = \sqrt{\frac{4\pi}{2L+1}} Y^*_{LM_L}(\hat{\mathbf{q}}) . \tag{B.25}$$

This property allows one to finally obtain the expression,

$$\rho(\mathbf{q}) = \sum_L 4\pi i^L Y^*_{LM_L}(\mathbf{q}) C_{LM_L}(q) , \tag{B.26}$$

which agrees with the expression in Eq. B.7.

Next, consider the Fourier transform of the longitudinal current,

$$\mathbf{j}_0(\mathbf{q}) = \int d\mathbf{x} e^{i\mathbf{q}\cdot\mathbf{x}} \hat{\mathbf{e}}_0 \cdot \mathbf{j}(\mathbf{x}) . \tag{B.27}$$

Once again, one may expand the exponential in partial waves to write,

$$\mathbf{j}_0(\mathbf{q}) = \int d\mathbf{x} \sum_J i^J \sqrt{4\pi(2J+1)} j_J(qx) Y_{J0}(\hat{\mathbf{x}}) \hat{\mathbf{e}}_0 \cdot \mathbf{j}(\mathbf{x}) . \tag{B.28}$$

Using the definition of vector spherical harmonics $\mathbf{Y}^M_{LJ1}$, it is possible to write [2]

$$Y_{J0}(\hat{\mathbf{x}}) \hat{\mathbf{e}}_\mu = \sum_{LM} \langle J0; 1\mu | LM \rangle \mathbf{Y}^M_{LJ1} . \tag{B.29}$$

Thus, the one can write

$$\mathbf{j}_0(\mathbf{q}) = \int d\mathbf{x} \sum_{JLM} i^J \sqrt{4\pi(2J+1)} j_J(qx) \langle J0;\, 10|LM\rangle \mathbf{Y}^M_{LJ1} \cdot \mathbf{j}(\mathbf{x})\,. \tag{B.30}$$

Properties of the Clebsch-Gordan coefficients restrict the sum to $M = 0$,

$$\mathbf{j}_0(\mathbf{q}) = \int d\mathbf{x} \sum_{JL} i^J \sqrt{4\pi(2J+1)} j_J(qx) \langle J0;\, 10|L0\rangle \mathbf{Y}^0_{LJ1} \cdot \mathbf{j}(\mathbf{x})\,, \tag{B.31}$$

and the triangle rule enforces the condition that $J = L, L \pm 1$. The relevant Clebsch-Grodan coefficient are,

$$\langle (L-1)0;\, 10|L0\rangle = \sqrt{\frac{L}{2L-1}}, \tag{B.32}$$

$$\langle L0;\, 10|L0\rangle = 0, \tag{B.33}$$

$$\langle (L+1)0;\, 10|L0\rangle = -\sqrt{\frac{L+1}{2L+3}}, \tag{B.34}$$

which gives the following expression,

$$\mathbf{j}_0(\mathbf{q}) = \int d\mathbf{x} \sum_{L} i^L \sqrt{4\pi} \left\{ -i\sqrt{L}\, j_{L-1}(qx) \mathbf{Y}^0_{L,L-1,1} - i\sqrt{L+1}\, j_{L+1}(qx) \mathbf{Y}^0_{L,L+1,1} \right\} \cdot \mathbf{j}(\mathbf{x})\,. \tag{B.35}$$

Employing the gradient formula for spherical harmonics [2] and properties of spherical Bessel functions, it is possible to show that [1]

$$\nabla[j_J(r) Y_{JM}] = \sqrt{\frac{J+1}{2J+1}} j_{J+1}(r) \mathbf{Y}^M_{J,J+1,1} + \sqrt{\frac{J}{2J+1}} j_{J-1}(r) \mathbf{Y}^M_{J,J-1,1}, \tag{B.36}$$

and therefore

$$\mathbf{j}_0(\mathbf{q}) = \int d\mathbf{x} \sum_{L} i^L \sqrt{4\pi(2L+1)} \left(-\frac{i}{q}\right) \nabla[j_L(qx) Y_{L0}] \cdot \mathbf{j}(\mathbf{x})\,. \tag{B.37}$$

Comparing with Eq. B.2, this is clearly

$$\mathbf{j}_0(\mathbf{q}) = -\sum_{L} i^L \sqrt{4\pi(2L+1)} L_{L0}(q)\,, \tag{B.38}$$

and rotating to an arbitrary direction following the same procedure as for $\rho(\mathbf{q})$, we obtain

$$\mathrm{j}_0(\mathbf{q}) = -\sum_L 4\pi i^L Y^*_{LM_L}(\hat{\mathbf{q}}) L_{LM_L}(q)\,, \tag{B.39}$$

which is consistent with Eq. B.8.

The final case to consider is the Fourier transform the current components transverse to the momentum transfer,

$$\mathrm{j}_\lambda(\mathbf{q}) = \int d\mathbf{x} e^{i\mathbf{q}\cdot\mathbf{x}} \hat{\mathbf{e}}_\lambda \cdot \mathbf{j}(\mathbf{x})\,, \tag{B.40}$$

where $\lambda = \pm 1$. Following a similar procedure as for the longitudinal current, one arrives at the expression,

$$\mathrm{j}_\lambda(\mathbf{q}) = \int d\mathbf{x} \sum_{JL} i^J \sqrt{4\pi(2J+1)}\, j_J(qx) \langle J0; 1\lambda | L\lambda\rangle \mathbf{Y}^\lambda_{LJ1} \cdot \mathbf{j}(\mathbf{x})\,. \tag{B.41}$$

Once again, the triangle rule restricts the values of J. The relevant Clebsch-Gordan coefficients in this case are,

$$\langle (L-1)0; 1\lambda | L\lambda\rangle = \sqrt{\frac{L+1}{2(L-1)}}, \tag{B.42}$$

$$\langle L0; 1\lambda | L\lambda\rangle = -\frac{\lambda}{\sqrt{2}}, \tag{B.43}$$

$$\langle (L+1)0; 1\lambda | L\lambda\rangle = \sqrt{\frac{L}{2(2L+3)}}\,, \tag{B.44}$$

Inserting these values into Eq. B.41, one arrives at the following result

$$\begin{aligned} \mathrm{j}_\lambda(\mathbf{q}) = \int d\mathbf{x} \sum_L i^L \sqrt{2\pi(2L+1)}\, \times \\ \left\{ -\lambda j_L(qx) \mathbf{Y}^\lambda_{LL1} - i \left[\sqrt{\frac{L+1}{2L+1}} j_{L-1}(qx) \mathbf{Y}^\lambda_{L,L+1,1} - \right.\right. \\ \left.\left. \sqrt{\frac{L}{2L+1}} j_{L+1}(qx) \mathbf{Y}^\lambda_{L,L-1,1} \right] \right\} \cdot \mathbf{j}(\mathbf{x})\,. \end{aligned} \tag{B.45}$$

Using the curl formula for spherical harmonics [2] and properties of the spherical Bessel functions, one finds that [1]

$$\mathbf{j}_\lambda(\mathbf{q}) = -\int d\mathbf{x} \sum_L i^L \sqrt{2\pi(2L+1)} \left[\frac{1}{q} \nabla \times j_L(qx)\mathbf{Y}^\lambda_{LL1} + \lambda j_L(qx)\mathbf{Y}^\lambda_{LL1} \right], \tag{B.46}$$

which, by inspection of Eqs. B.3 and B.4, yields

$$\mathbf{j}_\lambda(\mathbf{q}) = -\sum_L i^L \sqrt{2\pi(2L+1)}[E_{L\lambda}(q) + \lambda M_{L\lambda}(q)]. \tag{B.47}$$

Finally, to obtain an expression consistent with Eq. B.9, one may rotate to an arbitrary $\mathbf{q}$, which yields,

$$\mathbf{j}_\lambda(\mathbf{q}) = -\sum_{LM_L} i^L \sqrt{2\pi(2L+1)} D^L_{M_L,\lambda}(-\phi, -\theta, \phi)[E_{LM_L}(q) + \lambda M_{LM_L}(q)]. \tag{B.48}$$

In order to see the utility of the expressions of the charge and current matrix elements in terms of reduced multipoles, it useful to consider the generic expression for the interaction of a nucleus with an external field. We may consider the nuclear four-current $j_\mu(\mathbf{x})$ interacting with an external field $j^{\text{ext}}_\mu(\mathbf{x})$ transferring momentum $\mathbf{q}$ which, for some initial and final states $|i\rangle$ and $|f\rangle$, respectively, has the matrix elements,

$$\langle f|j^{\text{ext}}_\mu(\mathbf{x})|f\rangle = j^{\text{ext}}_\mu e^{-i\mathbf{q}\cdot\mathbf{x}}. \tag{B.49}$$

The external field could be a quantized field of radiation or the weak or electromagnetic interaction between the nucleus and an impinging lepton. The specific process will determine the details of $\mathbf{j}^{\text{ext}}_\mu(\mathbf{x})$ and the matrix elements in Eq. B.49. In general, a generic interaction Hamiltonians will have the form

$$H_{\text{int}} = C \int d\mathbf{x} j^{\text{ext}}_\mu(\mathbf{x}) j^\mu(\mathbf{x}), \tag{B.50}$$

where C is a generic coupling constant that will depend on the specific process considered. H_{int} has matrix elements[1, 4],

$$\langle f|H_{\text{int}}|f\rangle = C j^{\text{ext}}_\mu \int d\mathbf{x} e^{i\mathbf{q}\cdot\mathbf{x}} \langle -\mathbf{q};\, J_f M_f | j^\mu(\mathbf{x}) | \mathbf{J_i M_i}\rangle, \tag{B.51}$$

for initial and final nuclear states described by their angular momentum and projection on the spin quantization axis. Expanding out the contraction of the current operators lets one write Eq. B.51 as follows

$$\langle f|H_{\text{int}}|f\rangle = C \int d\mathbf{x} e^{i\mathbf{q}\cdot\mathbf{x}} \langle -\mathbf{q};\, J_f M_f | j^{\text{ext}}_0 \rho^\dagger(\mathbf{x}) - \mathbf{j}^{\text{ext}} \cdot \mathbf{j}^\dagger(\mathbf{x}) | \mathbf{J_i M_i}\rangle, \tag{B.52}$$

where,

$$\mathbf{j}^{\text{ext}} = \sum_{\lambda=0,\pm1} j^{\text{ext}}_\lambda \hat{\mathbf{e}}_\lambda. \tag{B.53}$$

Noting the following properties [2],

$$Y^*_{LM}(\mathbf{q}) = (-1)^M Y_{L,-M}(\mathbf{q}) \tag{B.54}$$

$$\mathbf{Y}^M_{LL1} = (-1)^{M+1}\mathbf{Y}^{-M}_{LL1}, \tag{B.55}$$

and choosing $M_i = M_f = M$ and employing Eq. B.14 we may obtain the following relations [4]

$$\langle J_f M|\rho^\dagger(q)|J_f M\rangle = (-1)^{J_i-M}\sum_L \sqrt{4\pi}(-i)^L P_L(\cos\theta)c^M_{J_f J_i L}C_L(q) \tag{B.56}$$

$$\langle J_f M|\hat{\mathbf{e}}^*_0\cdot\mathbf{j}^\dagger(q)|J_f M\rangle = (-1)^{J_i-M}\sum_L \sqrt{4\pi}(-i)^L P_L(\cos\theta)c^M_{J_f J_i L}L_L(q) \tag{B.57}$$

$$\begin{aligned}\langle J_f M|\hat{\mathbf{e}}^*_\lambda\cdot\mathbf{j}^\dagger(q)|J_f M\rangle &= (-1)^{J_i-M+1}\times\\ &\sum_{L\geq 1}\sqrt{8\pi^2}\frac{(-i)^L}{\sqrt{2L+1}}Y^*_{LM}(\theta,\phi)c^M_{J_f J_i L}\\ &\times[E_L(q)+M_L(q)]\end{aligned} \tag{B.58}$$

where $c^M_{J_f J_i L} = \langle J_f, M; J_i, -M|L, 0\rangle$, $P_L(\cos\theta)$ is a Legendre polynomial, and the following definition was employed

$$D^L_{0,\lambda}(-\phi,-\theta,\phi) = -\sqrt{\frac{4\pi}{2L+1}}Y_{L\lambda}(\theta,\phi), \tag{B.59}$$

which is true for $\lambda = \pm 1$. It is more convenient to recast Eq. B.58 in terms of the Cartesian components rather than the spherical components. Using the associated Legendre polynomial $P^1_L(x)$ defined for $L \geq 1$ as

$$P^1_L(x) = \sqrt{1-x^2}\frac{dP_L(x)}{dx}, \tag{B.60}$$

and noting that

$$\hat{\mathbf{e}}^*_\lambda = (-1)^\lambda\hat{\mathbf{e}}_\lambda, \tag{B.61}$$

we may convert Eq. B.58 into [4]

$$\begin{aligned}\langle J_f M|\hat{\mathbf{e}}_1\cdot\mathbf{j}^\dagger(q)|J_f M\rangle &= (-1)^{J_i-M+1}\sum_{L\geq 1}\sqrt{4\pi}(-i)^L c^M_{J_f J_i L}\frac{P^1_L(\cos\theta)}{\sqrt{L(L+1)}}\times\\ &[E_L(q)\cos\phi - iM_L(q)\sin\phi]\end{aligned} \tag{B.62}$$

$$\begin{aligned}\langle J_f M|\hat{\mathbf{e}}_2\cdot\mathbf{j}^\dagger(q)|J_f M\rangle &= (-1)^{J_i-M+1}\sum_{L\geq 1}\sqrt{4\pi}(-i)^L c^M_{J_f J_i L}\frac{P^1_L(\cos\theta)}{\sqrt{L(L+1)}}\times\\ &[E_L(q)\sin\phi + iM_L(q)\cos\phi].\end{aligned} \tag{B.63}$$

Thus for a suitable choice of **q** (and therefore of θ and ϕ), it is possible to extract all four of the standard reduced multipoles.

B.1 β-Decay Multipoles

The interaction Hamiltonian that mediates weak processes is [1]

$$H_W = -\frac{G_F V_{ud}}{\sqrt{2}} \int d\mathbf{x} j_\mu^{\text{lept}} \mathbf{j}^\mu(\mathbf{x})\,, \tag{B.64}$$

where the lepton current depends on the particular decay under study. The lepton current is well-understood and various expressions may be found in textbooks, such as Ref. [1]. For now, I will assume only that the current satisfies the condition in Eq. B.53. Namely,

$$j_\mu^{\text{lept}}(\mathbf{x}) = l_\mu e^{-\mathbf{q}\cdot\mathbf{x}}\,, \tag{B.65}$$

and from Eqs. B.7 through B.9, it may be shown that [1],

$$\begin{aligned}\langle f|H_W|f\rangle = -\frac{G_F V_{ud}}{\sqrt{2}} \langle f| \Bigg\{ &-\sum_{L\geq 1} \sqrt{2\pi(2L+1)}(-i)^L \sum_{\lambda=\pm 1} l_\lambda [\lambda M_{L,-\lambda}(q) + E_{L,-\lambda}(q)] \\ &+ \sum_{L\geq 0} \sqrt{4\pi(2L+1)}(-i)^L [l_3 L_{L0}(q) - l_0 M_{J0}(q)] \Bigg\} |i\rangle\,.\end{aligned} \tag{B.66}$$

Finally, using the Wigner-Eckart theorem (Eq. B.14) and the orthogonality of the Clebsch-Gordan coefficients, it is possible to obtain the square of Eq. B.66 averaged over the polarizations of the initial nucleus and summed over final state projections. This, of course, is necessary for Golden rule analyses of decay rates and cross sections. The end result is [1],

$$\begin{aligned}\frac{1}{2J_i+1}\sum_{M_f}\sum_{M_i} |\langle f|H_W|f\rangle|^2 = \frac{G_F V_{ud}^2}{\sqrt{2}} \Bigg\{ &\sum_{L\geq 1} \Bigg[\frac{1}{2}(\mathbf{l}\cdot\mathbf{l}^* - l_3 l_3^*)(|M_L(q)|^2 + |E_L(q)|^2) \\ &-\frac{i}{2}(\mathbf{l}\times\mathbf{l}^*)(2\text{Re}[M_L(q)E_L^*(q)] \Bigg] \\ &+ \sum_{L\geq 0} \Big[l_3 l_3^* |L_L(q)|^2 + l_0 l_0^* |C_L(q)|^2 \\ &-2\text{Re}(l_0 l_0^* L_L(q) C_L^*(q))\Big]\Bigg\}\,,\end{aligned} \tag{B.67}$$

which depends only on the reduced multipoles. Equation B.67 greatly simplifies in the limit $q \to 0$, which is a standard approximation for β-decay. Noting that as $qx \to 0$, the spherical Bessel functions behave like

$$j_L(qx) \to \frac{(qx)^L}{(2L+1)!!}, \tag{B.68}$$

the multipoles have the following low-momentum behavior [1]

$$C_{LM}(q) \approx \frac{q^L}{(2L+1)!!} \int d\mathbf{x} x^L Y_{LM} \rho(\mathbf{x}) \tag{B.69}$$

$$L_{LM}(q) \approx \frac{-iq^{(L-1)}}{(2L+1)!!} \int d\mathbf{x} x^L Y_{LM} \nabla \cdot \mathbf{j}(\mathbf{x}) \tag{B.70}$$

$$E_{LM}(q) \approx \frac{-iq^{(L-1)}}{(2L+1)!!} \sqrt{\frac{L+1}{L}} \int d\mathbf{x} x^L Y_{LM} \nabla \cdot \mathbf{j}(\mathbf{x}) \tag{B.71}$$

$$M_{LM}(q) \approx \frac{iq^L}{(2L+1)!!} \sqrt{\frac{L+1}{L}} \int d\mathbf{x} \left[\frac{1}{L+1} \mathbf{x} \times \mathbf{j}(\mathbf{x}) \right] \cdot \nabla x^L Y_{LM}, \tag{B.72}$$

with one exception in the case of,

$$L_{00}(q) \approx \frac{iq}{6} \int d\mathbf{x} x^2 Y_{00} \nabla \cdot \mathbf{j}(\mathbf{x}). \tag{B.73}$$

It is clear that, in the limit $q \to 0$, the only surviving multipoles are,

$$E_{1M}(q) = \sqrt{2} L_{1M}(q) = -\frac{i\sqrt{2}}{3} \sqrt{\frac{3}{4\pi}} \int d\mathbf{x} x Y_{1M} \nabla \cdot \mathbf{j}(\mathbf{x}) \tag{B.74}$$

$$C_{00}(q) = \sqrt{\frac{1}{4\pi}} \int d\mathbf{x} \rho(\mathbf{x}). \tag{B.75}$$

Parity selection rules imply that only the matrix elements involving the vector charge and axial three-current survive when $q \to 0$. Thus, Eq. B.74 is mediated by a purely axial operator and decays exhibiting these selection rules are referred to as "Gamow-Teller" transitions. Instead, the operator in Eq. B.75 is mediated by a scalar operator and this type of β-decay is known as "Fermi" decay.

B.2 The ^{6}He β-Decay Spectrum

To compute the β-decay spectrum with the standard multipole analysis on uses Eq. B.67 and the appropriate lepton tensor. Taking the expresion of the lepton current as in Ref. [1], it is possible to write the matrix element entering the differential decay rate $d\Gamma$ as,

$$
\begin{aligned}
\frac{1}{2J_i+1}\sum_{s_e,s_\nu}\sum_{M_i M_f}|\langle f|H_W|f\rangle|^2 = \frac{1}{2J_i+1}\sum_{J=0}^{\infty}\Big[&(1+\hat{\mathbf{k}}_\nu\cdot\hat{\mathbf{k}}_e)|C_J(q)|^2 \\
&+(1-\hat{\mathbf{k}}_\nu\cdot\hat{\mathbf{k}}_e+2(\hat{\mathbf{k}}_\nu\cdot\hat{\mathbf{q}})(\hat{\mathbf{q}}\cdot\hat{\mathbf{k}}_e))|L_J(q)|^2 \\
&-\hat{\mathbf{q}}\cdot(\hat{\mathbf{k}}_\nu+\hat{\mathbf{k}}_e)2\mathrm{Re}(L_J(q)C_J^*(q))\Big] \\
&+\sum_{J=1}^{\infty}\Big[(1-(\hat{\mathbf{k}}_\nu\cdot\hat{\mathbf{q}})(\hat{\mathbf{q}}\cdot\hat{\mathbf{k}}_e))(|M_J(q)|^2+|E_J(q)|^2) \\
&+\hat{\mathbf{q}}\cdot(\hat{\mathbf{k}}_\nu-\hat{\mathbf{k}}_e)2\mathrm{Re}(M_J(q)E_J^*(q))\Big],
\end{aligned} \tag{B.76}
$$

where s_e and s_ν are, respectively the electron and neutrino spins. The direction of neutrino momentum is $\hat{\mathbf{k}}_\nu = \mathbf{k}_\nu/E_\nu$, where E_ν is the neutrino energy, and direction of the electron momentum is $\hat{\mathbf{k}}_e = \mathbf{k}_e/\sqrt{k_e^2+m_e^2}$. It is also possible to link the momentum transfer to the lepton momenta through $\mathbf{q} = \mathbf{k}_e + \mathbf{k}_\nu$.

It is now worth noting the particular selection rules associated with this decay. Firstly, because the parity of the initial and final states are equal, the decay must be mediated by an operator with positive parity. Second, the allowed multipolarities in the decay rate are $J = |J_i - J_f| = 0, 1$. Together, this restricts the decay rate to four allowed multipoles: $C_1(q; A)$, $L_1(q; A)$, $E_1(q; A)$, and $M_1(q; V)$ where A or V indicates a multipole evaluated with only the axial or vector current, respectively. Then, considering Fermi's Golden Rule,

$$
d\Gamma = (2\pi)\delta(E_i - E_f - E_\nu - E_e)\frac{1}{2J_i+1}\sum_{\text{spins}}|\langle f|H_W|f\rangle|^2\frac{d\mathbf{k}_e}{(2\pi)^3}\frac{d\mathbf{k}_\nu}{(2\pi)^3}, \tag{B.77}
$$

the differential rate becomes,

$$
\begin{aligned}
d\Gamma = {} & (2\pi)\delta(E_i - E_f - E_\nu - E_e)\frac{1}{2J_i+1}G_F^2 V_{ud}^2\frac{4\pi}{2J_i+1} \\
&\Big[(1+\hat{\mathbf{k}}_\nu\cdot\hat{\mathbf{k}}_e)|C_1(q)|^2 + (1-\hat{\mathbf{k}}_\nu\cdot\hat{\mathbf{k}}_e+2(\hat{\mathbf{k}}_\nu\cdot\hat{\mathbf{q}})(\hat{\mathbf{q}}\cdot\hat{\mathbf{k}}_e))|L_1(q;A)|^2 \\
&-\hat{\mathbf{q}}\cdot(\hat{\mathbf{k}}_\nu+\hat{\mathbf{k}}_e)2\mathrm{Re}(L_1(q;A)C_1^*(q;a))\Big] + \Big[(1-(\hat{\mathbf{k}}_\nu\cdot\hat{\mathbf{q}})(\hat{\mathbf{q}}\cdot\hat{\mathbf{k}}_e))(|M_1(q;V)|^2+|E_1(q;A)|^2) \\
&+\hat{\mathbf{q}}\cdot(\hat{\mathbf{k}}_\nu-\hat{\mathbf{k}}_e)2\mathrm{Re}(M_1(q;V)E_1^*(q;A))\Big].
\end{aligned} \tag{B.78}
$$

B.3 ^{6}Li Muon Capture Rate

Muon capture on the atomic nucleus is a process in which a muon of mass m_μ with momentum $\mathbf{k}_\mu$ and energy E_μ and a proton in the nucleus of mass m_p are converted into a neutron of mass m_n and a neutrino of mass $m_\nu \simeq 0$ with momentum $\mathbf{k}_\nu$ and energy $E_\nu \simeq |\mathbf{k}_\nu|$. From Golden Rule considerations, the general expression for the capture rate (Γ), summed over the final states and averaged over the initial states, is given (in the limit of vanishing $\mathbf{k}_\mu$) by [1, 5]

$$d\Gamma = \frac{1}{2(2J_i+1)} \sum_{s_\nu, s_\mu, M_i, M_f} 2\pi\,\delta(\omega) |\langle \mathbf{k}_\nu, s_\nu;\, f, J_f M_f, -\mathbf{k}_\nu | H_W | \mathbf{k}_\nu, s_\nu;\, f, J_f M_f, -\mathbf{k}_\nu \rangle|^2 \frac{d^3 k_\nu}{(2\pi)^3}\,, \tag{B.79}$$

where the argument of the δ-function is

$$\omega = E_\nu + \sqrt{E_\nu^2 + (m_f + E_f)^2} - \left(m_\mu + m_i + E_i\right)\,, \tag{B.80}$$

and E_i and E_f are the initial and final state energies of the nucleus. This derivatoin neglects internal electronic energies that are on the order of tens of eV's for the light atoms under consideration. I also made use of the following definitions

$$\begin{aligned} m_i &= Z\,m_p + N\,m_n + (Z-1)\,m_e\,, \\ m_f &= (Z-1)(m_p + m_e) + (N+1)\,m_n\,, \end{aligned} \tag{B.81}$$

for an initial nucleus with charge number Z and neutron number N. The constants m_p, m_n, and m_e denote the usual proton, neutron, and electron masses, respectively. For the leptonic current, one may write,

$$\tilde{l}_\sigma(\mathbf{x}) = \bar{u}(\mathbf{k}_\nu, s_\nu)\gamma_\sigma(1-\gamma_5)\psi(x)u(\mathbf{k}_\mu, s_\mu)\,, \tag{B.82}$$

where the u's are spinors normalized to $u^\dagger u = 1$, the γ's are Bjorken-Drell convention gamma matrices, and s_ν and s_μ are the neutrino helicity and muon spin. $\psi(x)$ is the atomic wave function of the muon, which is taken to be a Hydrogen-like 1s wave function,

$$\psi_{1s}(x) = \sqrt{\frac{(Z\alpha\mu)^3}{\pi}} e^{-Z\alpha\mu x}\,, \tag{B.83}$$

where μ is the muon reduced mass and α is the fine structure constant. Replacing the two component spin state of the muon is justified in the limit $\mathbf{k}_\mu \to 0$. Because muon capture will be sensitive to the interior of the muonic wave function, one may further approximate the lepton tensor as $\tilde{l}_\sigma(\mathbf{x}) \simeq \psi_{1s}^{\mathrm{av}} l_\sigma(\mathbf{x})$ where ψ_{1s}^{av} is the average value of the wave function over the nuclear charge distribution [6],

$$\psi_{1s}^{\mathrm{av}} = \frac{\int d\mathbf{x} e^{i\mathbf{q}\cdot\mathbf{x}} \psi_{1s}(x) \rho_C(x)}{\int d\mathbf{x} e^{i\mathbf{q}\cdot\mathbf{x}} \rho_C(x)}\,, \tag{B.84}$$

with $\rho_C(x)$ being the electric charge density of the nucleus calculated from the many-body Hamiltonian. It is typical to write this as,

$$\mathcal{R} = \frac{|\psi_{1s}^{\mathrm{av}}|^2}{|\psi_{1s}(0)|^2}\,. \tag{B.85}$$

The properties of the delta function make the integration over E_ν rather simple. In the standard multipole analysis, noting that the parity and angular momentum selection rules are the same as in Sect. B.2, the partial rate to a single final state rate becomes [1]

$$\Gamma = \frac{G_F^2 V_{ud}^2}{2\pi} \frac{4\pi}{2J_i + 1} \mathcal{R} |\psi_{1s}(0)|^2 \frac{E_\nu^{*2}}{\text{recoil}} \times \left[|C_1(E_\nu^*; A) - L_1(E_\nu^*; A)|^2 + |E_1(E_\nu^*; A) - M_1(E_\nu^*; V)|^2 \right], \tag{B.86}$$

where the outgoing neutrino energy is given by energy conservation as [6]

$$E_\nu^* = \frac{\left(m_i + E_i + m_\mu\right)^2 - \left(m_f + E_f\right)^2}{2\left(m_i + E_i + m_\mu\right)}, \tag{B.87}$$

and the recoil factor arising from delta function properties is,

$$\frac{1}{\text{recoil}} = \left(1 - \frac{E_\nu^*}{m_i + E_i + m_\mu}\right). \tag{B.88}$$

In some cases, it is more convenient to cast this rate in terms of matrix elements. Making the choice $\hat{\mathbf{k}}_\nu = -\hat{\mathbf{q}} = -\hat{\mathbf{z}}$, one may write the rate as,

$$\begin{aligned}\Gamma = \frac{G_F^2 V_{ud}^2}{2\pi} \mathcal{R} \frac{|\psi_{1s}(0)|^2}{(2J_i + 1)} \frac{E_\nu^{*2}}{\text{recoil}} \times \\ \sum_{M_f, M_i} \Big[|\langle J_f, M_f | \rho(E_\nu^* \hat{\mathbf{z}}) | J_f, M_f \rangle|^2 + |\langle J_f, M_f | j_z(E_\nu^* \hat{\mathbf{z}}) | J_f, M_f \rangle|^2 \\ +2\mathrm{Re}\left[\langle J_f, M_f | \rho(E_\nu^* \hat{\mathbf{z}}) | J_f, M_f \rangle \langle J_f, M_f | j_z(E_\nu^* \hat{\mathbf{z}}) | J_f, M_f \rangle^* \right] \\ +|\langle J_f, M_f | j_x(E_\nu^* \hat{\mathbf{z}}) | J_f, M_f \rangle|^2 + |\langle J_f, M_f | j_y(E_\nu^* \hat{\mathbf{z}}) | J_f, M_f \rangle|^2 \\ -2\,\mathrm{Im}\left[\langle J_f, M_f | j_x(E_\nu^* \hat{\mathbf{z}}) | J_f, M_f \rangle \langle J_f, M_f | j_y(E_\nu^* \hat{\mathbf{z}}) | J_f, M_f \rangle^* \right] \Big]. \end{aligned} \tag{B.89}$$

References

1. J. D. Walecka. *Theoretical Nuclear and Subnuclear Physics*. Oxford University Press, New York, 1995.
2. Alan Robert Edmonds. *Angular momentum in quantum mechanics*, volume 4. Princeton university press, 1996.
3. J. Carlson and R. Schiavilla. Structure and dynamics of few nucleon systems. *Rev. Mod. Phys.*, 70:743–842, 1998. doi: 10.1103/RevModPhys.70.743.
4. R Schiavilla. Weak transitions in nuclei. unpublished, 2015.

5. L. E. Marcucci, R. Schiavilla, M. Viviani, A. Kievsky, S. Rosati, and J. F. Beacom. Weak proton capture on ^{3}He. *Phys. Rev. C*, 63:015801, Dec 2000. doi: 10.1103/PhysRevC.63.015801. URL https://link.aps.org/doi/10.1103/PhysRevC.63.015801.
6. L. E. Marcucci, M. Piarulli, M. Viviani, L. Girlanda, A. Kievsky, S. Rosati, and R. Schiavilla. Muon capture on deuteron and ^{3}He. *Phys. Rev. C*, 83:014002, Jan 2011. doi: 10.1103/PhysRevC.83.014002. URL https://link.aps.org/doi/10.1103/PhysRevC.83.014002.

Appendix C
Radiative Corrections to ^{6}He β-Decay

In this appendix, I summarize the necessary Coulomb and kinematic radiative corrections adopted in Ref. [1] needed to compute the ^{6}He β-decay spectrum. This was first summarized in Appendix E of the aforementioned reference and the formalism is that of Ref. [2], which provides an in depth overview of the present status for including radiative corrections in β-decay calculations.

C.1 Coulomb and Radiative Corrections

Here I report on the radiative corrections to ^{6}He β-decay which were first discussed in Appendix E of Ref. [1]. Radiative corrections to β-decay may be divided into Coulomb corrections and so called "outer" and "inner" corrections. The latter two come from photon emissions processes not included in Coulomb corrections that are infrared divergent and kinematic-independent radiative corrections, respectively. Inner and outer corrections have been studied extensively [3, 4] and were simply inserted into the final result for the ^{6}He β-decay spectrum in Eq. 4.19. Coulomb corrections, instead, are of interest for the level of experimental precision currently pursued and to first order can be understood as the elastic response of the γW box diagrams. Higher-order ($\mathcal{O}\{[\alpha Z]^n\}$) behavior is then absorbed by substituting the electron plane wave by a solution of the Dirac equation in the static, spherical potential of the atomic final state, $\phi_e(\boldsymbol{r}, \boldsymbol{p}_e)$, and writing the Hamiltonian as

$$\mathcal{M}_{fi} = \int \mathrm{d}^3 r\, \bar{\phi}_e(\boldsymbol{r}, \boldsymbol{p}_e)\gamma^\mu(1-\gamma^5)v(\boldsymbol{p}_{\bar{\nu}}) \\ \times \int \frac{\mathrm{d}^3 s}{(2\pi)^3} e^{i\boldsymbol{s}\cdot\boldsymbol{r}} \frac{1}{2}[\langle f(\boldsymbol{p}_f + \boldsymbol{p}_e - \boldsymbol{s})|V_\mu + A_\mu|i(\boldsymbol{p}_i)\rangle + \langle f(\boldsymbol{p}_f)|V_\mu + A_\mu|i(\boldsymbol{p}_i - \boldsymbol{p}_e + \boldsymbol{s})\rangle]. \tag{C.1}$$

G. B. King, *Electroweak Structure and Reactions in Light Nuclei with Quantum Monte Carlo Methods*, Springer Theses, https://doi.org/10.1007/978-3-032-03191-4

The multipole expansion proceeds in the same way and the most immediate modification comes from the Fermi function given by Eq. 4.20. Additional small terms are introduced that modify the differential decay rate that are well-known in the literature and can be found in several references [5, 6]. Additional known correction terms arise when moving past a point-charge model of the nucleus and introduce additional subdominant electromagnetic corrections; for instance, screening by atomic electrons and atomic exchange processes. Reference [1] made use of the results of Ref. [2] by translating the multipole coefficients into the Behrens-Bühring formalism [7, 8], resulting in the following additional terms in the deecay rate,

$$C_{\text{Coulomb}}(Z,\varepsilon) = \alpha Z W_0 R \left[|L_1^{(0)}|^2 \left(\frac{6}{35} - \frac{233}{210}\frac{\alpha Z}{W_0 R} - \frac{3}{70}\frac{\mu_e^2}{\varepsilon} - \frac{12}{7}\varepsilon \right) + \left(\frac{r_\pi}{R}\right)^2 \left(\frac{E_1^{(0)}E_1^{(2)}}{2} - L_1^{(0)}L_1^{(2)} \right) \left(-\frac{20}{35} + \frac{4}{7}\varepsilon \right) \right] \tag{C.2}$$

where the fractional prefactors were calculated assuming a uniformly charged sphere as discussed in Ref. [2].

All together, the inclusion of electroweak corrections modifies the β-decay spectrum by

$$\frac{d\Gamma}{d\varepsilon} \propto (1+\Delta_R^V)(1+\delta_R(\varepsilon))F_0(Z,\varepsilon)L_0(Z,\varepsilon)S(Z,\varepsilon)[C_0(\varepsilon)+C_{\text{Coulomb}}(Z,\varepsilon)] \tag{C.3}$$

where Δ_R^V is the inner radiative corrections to vector transitions (the difference induced by axial transitions is small and absorbed into an experimentally determined g_A value) and $\delta_R(\varepsilon)$ are outer radiative corrections to $\mathcal{O}(\alpha^3 Z^2)$. Note that ε was defined in Sect. 4.3 such that $E_e = W_0\varepsilon$ for endpoint energy W_0 of the decay. The product $F_0 L_0$ is the Fermi function for a uniformly charged sphere, S describes the shielding of the nuclear charge by atomic electrons and $C_{\text{Coulomb}}(Z,\varepsilon)$ are the modifications due to higher-order Coulomb corrections folded with the nuclear current.

C.2 Kinematic Recoil Corrections

In Appendix B, I introduced the multipole expansion of the weak Hamiltonian [9]. This expansion is not Lorentz covariant and is implicitly performed in the Breit frame ($\boldsymbol{p}_i = -\boldsymbol{p}_f$). In the approximation of an infinitely massive nucleus, the Breit and lab frames agree. These results, however, must be Lorentz boosted back into the lab frame for a finite momentum transfer, leading to additional $\mathcal{O}(q/M)$ results as discussed in Ref. [8]. The kinematic recoil corrections originating from the phase space integral may also be treated as a multiplicative factor

$$d\Gamma \propto |\mathcal{M}|^2 \left(1 + \frac{3E_e - W_0 - 3\boldsymbol{p}_e \cdot \hat{\boldsymbol{v}}_\nu}{M}\right) \tag{C.4}$$

where the final term gives a finite contribution in the neutrino angular integral when combined with the β-ν correlation. Combining these terms leads to the total kinematic recoil corrections [2, 10]

$$R_N \approx 1 - \frac{2W_0}{3M} + \frac{10E_e}{3M} - \frac{2m_e^2}{3E_e M} \tag{C.5}$$

where here only terms to first order in q/M are retained. Neglected terms would lead to corrections of $\mathcal{O}(10^{-6})$, at most.

References

1. Garrett B. King, Alessandro Baroni, Vincenzo Cirigliano, Stefano Gandolfi, Leendert Hayen, Emanuele Mereghetti, Saori Pastore, and Maria Piarulli. Ab initio calculation of the β-decay spectrum of He6. *Phys. Rev. C*, 107(1):015503, 2023. doi: 10.1103/PhysRevC.107.015503.
2. Leendert Hayen, Nathal Severijns, Kazimierz Bodek, Dagmara Rozpedzik, and Xavier Mougeot. High precision analytical description of the allowed $$ spectrum shape. *Reviews of Modern Physics*, 90(1):015008, mar 2018. ISSN 0034-6861. doi: 10.1103/RevModPhys.90.015008. URL https://link.aps.org/doi/10.1103/RevModPhys.90.015008.
3. Alberto Sirlin and Andrea Ferroglia. Radiative corrections in precision electroweak physics: A historical perspective. *Reviews of Modern Physics*, 85(1):263–297, 2013. ISSN 00346861. doi: 10.1103/RevModPhys.85.263.
4. I. S. Towner and J. C. Hardy. Improved calculation of the isospin-symmetry-breaking corrections to superallowed Fermi β decay. *Physical Review C*, 77(2):025501, feb 2008. ISSN 0556-2813. doi: 10.1103/PhysRevC.77.025501. URL https://link.aps.org/doi/10.1103/PhysRevC.77.025501.
5. H. Behrens, H. Genz, M. Conze, H. Feldmeier, W. Stock, and A. Richter. Allowed β-transitions, weak magnetism and nuclear structure in light nuclei. *Annals of Physics*, 115(2):276–324, oct 1978. ISSN 00034916. doi: 10.1016/0003-4916(78)90158-6. URL http://linkinghub.elsevier.com/retrieve/pii/0003491678901586 https://linkinghub.elsevier.com/retrieve/pii/0003491678901586.
6. Barry R. Holstein. Recoil Effects in Allowed beta Decay: The Elementary Particle Approach. *Rev. Mod. Phys.*, 46:789, 1974. doi: 10.1103/RevModPhys.46.789. [Erratum: Rev.Mod.Phys. 48, 673–673 (1976)].
7. H Behrens and W Bühring. *Electron radial wave functions and nuclear beta-decay*. Clarendon Press, Oxford, 1982.
8. Leendert Hayen and Albert R. Young. Consistent description of angular correlations in β decay for Beyond Standard Model physics searches. 9 2020.
9. J. D. Walecka. *Theoretical Nuclear and Subnuclear Physics*. Oxford University Press, New York, 1995.
10. D.H. Wilkinson. Analysis of neutron $$-decay. *Nuclear Physics A*, 377(2-3):474–504, 1982. ISSN 03759474. doi: 10.1016/0375-9474(82)90051-3.

Appendix D
β-Decay Shapes

In this section, I present a simple example of how beyond the Standard Model (BSM) currents would impact β-decay shapes. I show how the angular correlations and spectrum are distorted in a simple, worked example. Emphasis is placed on how a tensor would impact a purely GT decay.

D.1 General Features of Differential Rate

For the β-decay of nucleus with angular momentum $J = 0$, the most general parity violating semi-leptonic Hamiltonian containing vector (V), scalar (S), axial-vector (A), tensor (T), and pseudoscalar interactions in the limit of no nuclear recoil has the differential rate [1],

$$d\Gamma = \frac{|M|^2}{2(2\pi)^5} kE(W_0 - E)^2 \left(1 + a\cos\theta + b\frac{m}{E}\right) dE d\Omega_e d\Omega_\nu \,, \tag{D.1}$$

where k and E are the electron momentum and energy, respectively, m is the electron mass, θ is the angle between the outgoing electron and neutrino momenta, $\Omega_{e(\nu)}$ is the electron (neutrino) solid angle, and W_0 is the energy released in the decay. The differential rate contains the semi-leptonic matrix element M, the angular correlation a, and the "Fierz" interference term b. These three quantities depend on eight coupling coefficients contained in the lepton current; however in this discussion, I will assume time-reversal invariance, which reduces the number to four. For the purposes of demonstrating the effect on the spectrum, I will also assume real coefficients, though in principle they may be complex. Under these assumptions, a and b reduce to,

G. B. King, *Electroweak Structure and Reactions in Light Nuclei with Quantum Monte Carlo Methods*, Springer Theses, https://doi.org/10.1007/978-3-032-03191-4

$$|M|^2 = |M_F|^2(C_V^2 + C_S^2) + |M_{GT}|^2(C_A^2 + C_T^2)\,, \tag{D.2}$$

$$a = \frac{1}{|M|^2}\left[|M_F|^2(C_V^2 - C_S^2) + \frac{|M_{GT}|^2}{3}(C_T^2 - C_A^2)\right]\,, \tag{D.3}$$

$$b = \frac{2}{|M|^2}\left[|M_F|^2 C_S C_V + |M_{GT}|^2 C_T C_A\right]\,, \tag{D.4}$$

were C_V, C_S, C_T, and C_A are the V, S, T, and A coupling coefficients, respectively, and the quantities M_F and M_{GT} are the Fermi (F) and Gamow-Teller (GT) matrix elements.

From Eqs. D.2 through D.4, there are two features which are immediately clear. The first is that if the F or GT transition is "pure," there is no Fierz term. This is the case for the $V - A$ theory of the standard model, where $C_V = 1$, $C_A = -1$, and $C_S = C_T = 0$. The second is that pure F or pure GT transitions in the zero recoil limit will have a well-defined value for a that does not depend on the nuclear matrix element. In the Standard Model, $a_F = 1$ and $a_{GT} = -1/3$. Recoil, of course, will modify the angular correlation even in these pure cases. In the limit of no recoil, a pure scalar interaction has $a = -1$ and a pure tensor results in $a = 1/3$.

D.2 Angular Distribution

One can consider the distribution of decays in $\cos\theta$ by fixing the electron angle and integrating over its energy. By change of variables, one can show that fixing the electron angle results in the transformation,

$$d\Omega_e d\Omega_\nu \rightarrow 8\pi^2 d(\cos\theta) \tag{D.5}$$

and thus the angular distribution may derived from Eq. D.1,

$$\frac{d\Gamma}{d(\cos\theta)} = G[1 + a\cos\theta]\,, \tag{D.6}$$

where G is a prefactor determined by the integration over energy which has no angular dependence, but is clearly sensitive to any beyond the Standard Model (BSM) interference with the SM; however, note that a subsequent integration over angle, independent of the value of a, results in,

$$\Gamma = 2G\,, \tag{D.7}$$

which makes it instructive to investigate,

$$\frac{1}{\Gamma}\frac{d\Gamma}{d(\cos\theta)} = \frac{1}{2}\left[1 + a\cos\theta\right]\,, \tag{D.8}$$

Fig. D.1 Probability distribution of lepton angle for pure axial (blue solid line), tensor (orange dashed line), vector (magenta dot-dashed line), and scalar (green dotted line) currents

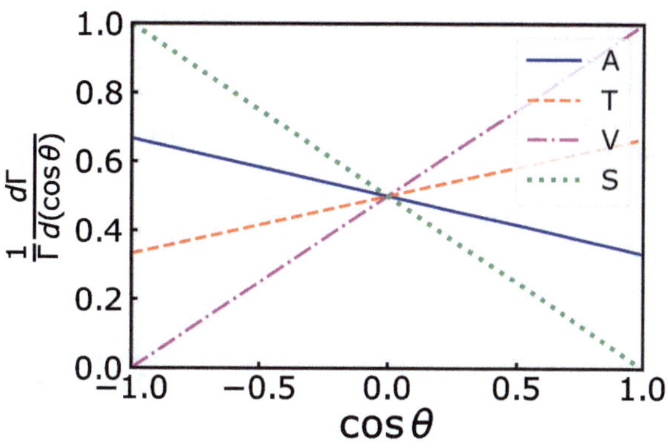

which gives the probability distribution for a decay having a given angle between the leptons. In Fig. D.1, I show the characteristic angular distribution for currents of a pure Lorentz character. Pure vector and tensor currents preferentially emit leptons with their momentum vectors parallel, while axial and scalar currents prefer the back-to-back configuration.

Now, I will focus on a decay that is GT-only and assume some correction from a BSM tensor. For this case, I define a as function of the ratio $\epsilon \equiv C_\mathrm{T}/C_\mathrm{A}$,

$$a(\epsilon) = -\frac{1}{3}\left[\frac{1-\epsilon^2}{1+\epsilon^2}\right], \tag{D.9}$$

Equation. D.9 shows that, independent of the sign of C_T, adding a tensor correction will increase the value of a. In Fig. D.3, I show the effect of an exaggeratedly large value $\epsilon = 0.1$ on the probability to emit back-to-back leptons. What the tensor does, in effect, is reduce that probability while increasing the probability of producing two parallel leptons (Fig. D.2).

Fig. D.2 Probability distribution of lepton angle for a pure axial current (blue solid line) and axial-plus-tensor current with $\epsilon = 0.1$ (orange dot dashed line) in the limit of no nuclear recoil

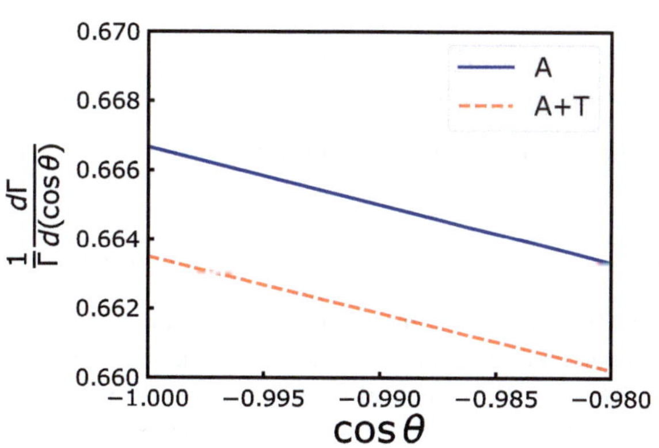

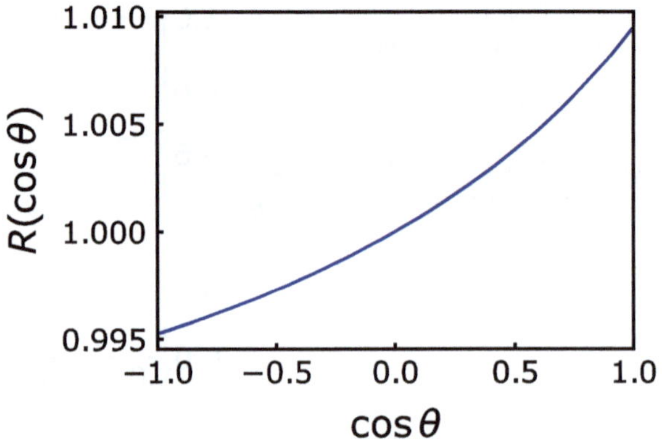

Fig. D.3 The ratio of the axial-plus-tensor angular distribution with $\epsilon = 0.1$ to the pure axial angular distribution

To further illuminate how the added tensor changes the distribution, I also consider the ratio,

$$R(\cos\theta) = \frac{1 + a(0.1)\cos\theta}{1 - \cos\theta/3}, \tag{D.10}$$

which shows how the angular distribution is distorted by the presence of the tensor. Here, one clearly sees the suppression of back-to-back leptons and an enhancement of parallel leptons.

D.3 Energy Distribution

While a tensor will always impact the angular distribution in a characteristic way independent of its sign relative to the axial coupling, the same is not true for the energy distribution. Here, I will consider only the purely GT case with BSM tensor interference. From Eq. D.1, the decay rate is

$$\frac{d\Gamma}{dE} = \frac{|M_{\mathrm{GT}}|^2(1+\epsilon^2)}{4\pi^3} kE(W_0 - E)^2 \left[1 + b(\epsilon)\frac{m}{E}\right], \tag{D.11}$$

where I have made the definition,

$$b(\epsilon) = \frac{2\epsilon}{1+\epsilon^2}. \tag{D.12}$$

Adding the tensor changes the overall prefactor and introduces a characteristic energy dependence that is sensitive to the relative sign between the axial and tensor contributions. For illustration, I will consider the spectra generated in the two cases $\epsilon = \pm 0.1$. For this discussion, I also take $M_{\mathrm{GT}} = 2$ and $W_0 = 4.5$ MeV, which are similar to the case for ^{6}He β-decay. I will plot the rate in natural units (where the spectrum is dimensionless) to compare the shapes of the distributions.

The absolute spectra for the adopted values of the parameters are shown in Fig. D.4. The positive value of ϵ enhances the rate over all, while the energy dependence for negative ϵ decreases the rate. As expected from the analysis in Ref. [2], there is not as much sensitivity to b at the ends of the spectrum as compared to the region from 1 to 2 MeV. The reason for this is two fold; first, at large E, the distortion decreases with E^{-1}. Secondly, the spectrum must vanish at m_e and W_0 from energy conservation, so it is practically difficult to extract b as you approach those limits.

To connect with the plots of Ref. [3], I also show the relative spectra in Fig. D.5 Indeed, there is a distortion that is larger at low E from the Fierz term and at large E, the ratios asymptotically approach $(1+\epsilon^2)$. Physically, one may interpret this behavior as, at large energies, the A and T contributions summing coherently. Because the electron chirality coincides with helicity when it becomes relativistic and the tensor flips the chirality relative to the axial current, this is a manifestation of the conservation of helicity preventing the two currents from interfering.

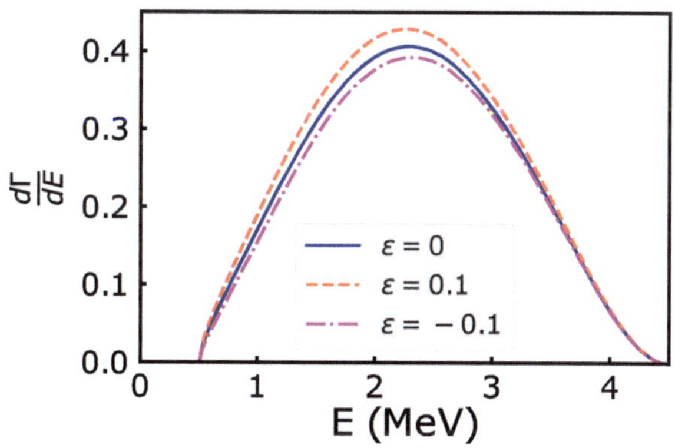

Fig. D.4 The absolute β-decay spectra in the no recoil limit for the Standard Model (solid blue) and with the inclusion of a BSM tensor current having either $\epsilon = 0.1$ (dashed orange) or $\epsilon = -0.1$ (dot-dashed magenta)

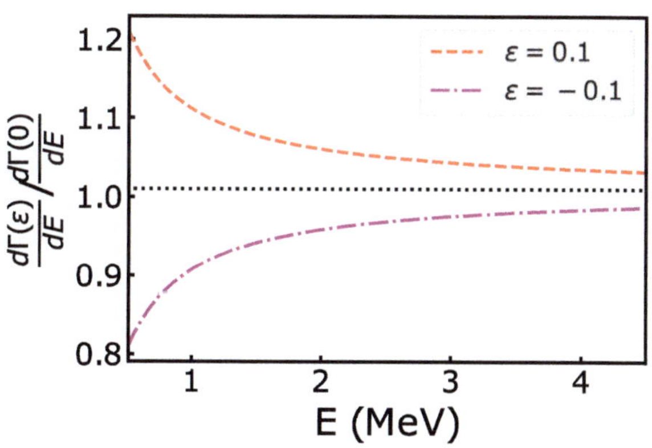

Fig. D.5 The β-decay spectra relative to the Standard model in the no recoil limit for the inclusion of a BSM tensor current having either $\epsilon = 0.1$ (dashed orange) or $\epsilon = -0.1$ (dot-dashed magenta). The dotted black line represents the asymptotic limit $(1 + c^2)$

References

1. J. D. Jackson, S. B. Treiman, and H. W. Wyld. Possible tests of time reversal invariance in Beta decay. *Phys. Rev.*, 106:517–521, 1957. doi: 10.1103/PhysRev.106.517.
2. Martin Gonzalez-Alonso and Oscar Naviliat-Cuncic. Kinematic sensitivity to the Fierz term of β-decay differential spectra. *Phys. Rev. C*, 94(3):035503, 2016. doi: 10.1103/PhysRevC.94.035503.
3. Garrett B. King, Alessandro Baroni, Vincenzo Cirigliano, Stefano Gandolfi, Leendert Hayen, Emanuele Mereghetti, Saori Pastore, and Maria Piarulli. Ab initio calculation of the β-decay spectrum of He6. *Phys. Rev. C*, 107(1):015503, 2023. doi: 10.1103/PhysRevC.107.015503.

Appendix E
Multipole Decomposition of Magnetic Form Factors

The following appendix follows from the derivation of magnetic form factors in Ref. [1]. In VMC and GFMC, the linear equations needed for the extraction of the reduced matrix element $M_1(q)$ needed for the magnetic form factor (and, indeed, for any reduced multipole) are conveniently obtained by changing the direction of momentum transfer $\hat{q}$ rather than repeating the calculation of the matrix elements for different projections of the total angular momentum. Therefore, if there are n possible multipoles, one should select n values for the angle θ between the $\hat{\mathbf{z}}$ axis and the direction $\hat{\mathbf{q}}$. This results in n linear equations to invert to determine the reduced multipoles. It does not matter which θ is chosen and thus three values were selected to give the simplest extraction. To simplify the expressions, the matrix element computed using Monte Carlo is denoted as

$$X_\theta^J = \left\langle J\ J \left| j_{\gamma,y}(q(\theta)) \right| J\ J \right\rangle , \tag{E.1}$$

and

$$M_L^J = \langle J J || M_L || J J \rangle , \tag{E.2}$$

represents the reduced matrix elements of the magnetic multipole L for angular eigenstate state J with projection on the spin-quantization axis J.

For nuclei with $J = 1/2$ and $J = 1$, only the multipole with $L = 1$ contributes to the form factor. In this case, only one equation is needed and $\theta = \pi/2$ ($\hat{\mathbf{q}} = \hat{\mathbf{x}}$) which gives the magnetic moment at $q = 0$ sufficies. The reduced matrix element for $J = 1/2$ and $J = 1$ is thus simply,

$$M_1^{1/2,1} = \frac{1}{\sqrt{\pi}} X_{\pi/2}^{1/2,1} . \tag{E.3}$$

G. B. King, *Electroweak Structure and Reactions in Light Nuclei with Quantum Monte Carlo Methods*, Springer Theses, https://doi.org/10.1007/978-3-032-03191-4

The situation for the $J = \frac{3}{2}$ case is now more complicated because both the M_1 and M_3 multipoles contribute to the form factor. In order to obtain the two magnetic multipoles, two independent calculations must be performed. Seclting $\theta = \pi/2$ and $\theta = \pi/4$ and inverting the linear equations gives the following expressions,

$$M_1^{3/2} = \sqrt{\frac{2}{5\pi}}\left[X_{\pi/2}^{3/2} + \frac{2\sqrt{2}}{3}X_{\pi/4}^{3/2}\right], \tag{E.4}$$

$$M_3^{3/2} = \frac{8}{\sqrt{15\pi}}\left[X_{\pi/2}^{3/2} - \sqrt{2}X_{\pi/4}^{3/2}\right]. \tag{E.5}$$

For the $J = 2$ case, it is again possible to excite only two multipoles. Using the same angles as for $J = 3/2$ gives,

$$M_1^2 = \frac{3}{2\sqrt{5\pi}}\left[X_{\pi/2}^2 + \frac{2\sqrt{2}}{3}X_{\pi/4}^2\right], \tag{E.6}$$

$$M_3^2 = \frac{4\sqrt{2}}{\sqrt{15\pi}}\left[X_{\pi/2}^2 - \sqrt{2}X_{\pi/4}^2\right]. \tag{E.7}$$

For $J = 3$, the multipolarities $L = 1,\ 3$, and 5 may be excited. Therefore, three independent linear equations are necessary. Choosing $\theta = \pi/2, \theta = \pi/4$ and $\cos\theta = 2\sqrt{2}/3$, the solution of the linear equations reads

$$M_1^3 = \sqrt{\frac{1}{7\pi}}\left[\frac{53}{280\sqrt{2}}X_{\pi/2}^3 + \frac{232}{735}X_{\pi/4}^3 - \frac{81}{1960\sqrt{2}}X_{1/3}^3\right], \tag{E.8}$$

$$M_3^3 = \frac{1}{\sqrt{\pi}}\left[\frac{13\sqrt{2}}{20}X_{\pi/2}^3 - \frac{32}{35}X_{\pi/4}^3 - \frac{81\sqrt{2}}{140}X_{1/3}^3\right], \tag{E.9}$$

$$M_5^3 = 6\sqrt{\frac{35}{\pi}}\left[\frac{\sqrt{2}}{35}X_{\pi/2}^3 - \frac{32}{245}X_{\pi/4}^3 + \frac{27\sqrt{2}}{245}X_{1/3}^3\right], \tag{E.10}$$

where $X_{1/3}^3 = \langle 3\,3|\, j_{\gamma,y}(q(\cos\theta = 1/3))\,|3\,3\rangle$. I stress that no assumption about the nucleus was made other than that it is an angular momentum eigenstate; therefore, these sets of equations could be used to extract the magnetic form factors of any nucleus with the appropriate J.

Reference

1. G. Chambers-Wall, A. Gnech, G. B. King, S. Pastore, M. Piarulli, R. Schiavilla, and R. B. Wiringa. Magnetic structure of A≤10 nuclei using the Norfolk nuclear models with quantum Monte Carlo methods. *Phys. Rev. C*, 110(5):054316, 2024. doi: 10.1103/PhysRevC.110.054316.

MIX
Papier aus verantwortungsvollen Quellen
Paper from responsible sources
FSC® C105338

If you have any concerns about our products,
you can contact us on
ProductSafety@springernature.com

In case Publisher is established outside the EU,
the EU authorized representative is:
Springer Nature Customer Service Center GmbH
Europaplatz 3, 69115 Heidelberg, Germany

Printed by Libri Plureos GmbH
in Hamburg, Germany